European Crossbows: A Survey by Josef Alm

Translated by H Bartlett Wells

Edited by G M Wilson

Royal Armouries Museum
Armouries Drive
Leeds LS10 1LT

© 1994 The Trustees of the Royal Armouries and
the Arms and Armour Society

All rights reserved. No part of this publication
may be reproduced, stored in a retrieval system,
or transmitted in any form by any means,
without first obtaining the written permission of
the copyright owner.

First published 1994, reprinted with amendments 1998.
Reprinted 2001.

ISBN 0 948092 20 3

Printed by Henry Ling Ltd, The Dorset Press,
Dorchester, Dorset.

Front cover illustration
A detail of one of the carved bone panels
covering the tiller of a mid 16th century
German sporting crossbow. The figure appears
to be carrying an arrow or crossbow bolt
and may represent a captain of archers.
Royal Armouries. XI.8.

Contents

Editor's Introduction	2
EUROPEAN CROSSBOWS: A SURVEY BY JOSEF ALM	
1 Origin and primitive European types	6
2 The 12th and 13th Centuries	14
3 The 14th Century	25
4 The 15th Century	33
5 The Crossbow after 1500: Central Europe	55
6 The Crossbow after 1500: Western Europe and Spain	64
7 The Crossbow after 1500: The Northern Countries	69
8 Pellet Crossbows	93
9 References	98
Bibliography	105

Introduction

Josef Alm, the Swedish arms historian, published his *Europeiska Armborst* in *Vaabenhistoriske Aarbøger* (*The Yearbook of the Danish Arms and Armour Society*) in 1947. It was then by far the best overall survey of the development of the European crossbow and even today, despite over 40 years of research and many publications on the subject, it remains a most important source of information for all those interested in crossbows.

Until now, however, this important work has not been accessible to most English speaking students of the subject as there was no published translation of the Swedish text apart from a very short German resumé printed at the end of the original text.

In 1972 the Arms and Armour Society of Great Britain took the initiative and paid for a translation of Alm's work by H Bartlett Wells, with the intention of publishing it immediately. The Society of Archer-Antiquaries undertook to help by editing the typescript and Commander W F Paterson, himself a considerable authority on crossbows, began work. Unfortunately, because of the amount of editing required, delays occurred and before it was completed the Arms and Armour Society was forced to postpone publication for financial reasons. Commander Paterson continued with some excellent editing and prepared a draft translation, some editorial notes and the outline of an index. He was assisted by Mr Finn Askgaard of the Tøjhusmuseum, Copenhagen. The text was also seen and worked on by Claude Blair and Ian Eaves for the Arms and Armour Society. I became involved in 1983 when the Arms and Armour Society asked me to look at the partly edited text and advise on whether it was still worthy of publication, and if so how it could be done. In 1986 it was finally agreed that I would undertake the still considerable task of preparing the translation for publication and that the Royal Armouries would publish it on behalf of the Society. It has taken seven years of work fitted into the occasional free hours in the busy life that is the lot of a Master of the Armouries today. The project has travelled around the world with me, and been worked on in planes and hotel bedrooms as well as in the quiet of my own study at home.

The editing has entailed more work than I originally estimated. Alm was a self-taught man and wrote in a rather idiosyncratic style. I have tried to preserve some of the flavour of this, while improving and modernising the English vocabulary and idiom of the original translation. I have attempted to translate accurately Swedish crossbow terminology into English words and phrases which students of crossbows will readily understand.

I have left the text much as Alm wrote it and have continued to use his unusual system of referencing. I have corrected only one mistake in the text itself, although other errors are mentioned in the editor's footnotes on each page. Where to improve understanding it has been necessary to insert something into the text this has always been done using square brackets. The editor's notes, I hope, will help to make the reader understand the text better. Alm, naturally, concentrated his researches on Scandinavia and Eastern Europe and his text is thus full of references to people, places and events that will mean little to many readers of this translation. In the notes I have attempted to give sufficient geographical, historical, and occasionally other types of information to help the reader understand what Alm was writing.

In addition, I have occasionally drawn attention to inadequacies in the text, and to advances in knowledge in particular areas. All the line drawings from the original publication have been photographed and are reproduced here. Prints of the same photographs as those used by Alm have been acquired and used wherever possible, and I am grateful to all the institutions involved for permission to reproduce them. Where this has proved impossible, either a similar alternative photograph of the same object has been used, or the plate from the original publication has been photographed and is reproduced together with a better photograph of a similar alternative object. It was decided at the beginning of my involvement with this project that an index would not be produced, and I hope that this does not cause too much irritation to serious researchers or general readers. The volume concludes with a comprehensive bibliography of books and articles on crossbows prepared by Sarah Barter Bailey, the Librarian of the Royal Armouries. It is, I think a fitting end piece to Josef Alm's considerable contribution to the study of the crossbow.

I have been helped by many people in the preparation of this volume. My thanks go to all those involved in preparing and editing the translation before it came to me: to the late Commander Paterson who did so much to encourage my interest in this subject; to the President and Council of the Arms and Armour Society for allowing me the privilege of being involved with this project; to Howard Blackmore and to my predecessor as Master of the Armouries, Nick Norman, for much helpful advice and encouragement; to the typists in the Royal Armouries, and especially to Denise Ferry, who patiently retyped the script many, many times; to Ole Frantzen, Director of the Tøjhusmuseum, Copenhagen, for helping me with Danish queries; and finally, and especially, to Nils Drejholt, of the Livrustkammare, Stockholm. Nils very graciously undertook to check the entire edited translation against the original to ensure its accuracy and has made many suggestions which have helped to improve the work. He has contributed, too, invaluable information about many subjects which I have used in the editor's notes to the text, and it was he who suggested that I include the brief introduction to Scandinavian history which follows. Without his efforts, and those of many others, this work would be much the poorer.

However, even with their help, there are doubtless mistakes and inconsistencies in these pages. For these I take full responsibility. I apologise for them and hope that they do not lead others into error or prevent readers from enjoying and appreciating the work of Josef Alm, which appears here in English for the first time. Finally, I gratefully acknowledge the kind permission to publish this translation given by the Armémuseum and Livrustkammaren, Stockholm, which museums hold the copyright in Alm's work.

Josef Alm (1889-1957)

By the time of his death in 1957 Josef Alm had become widely respected internationally as an arms and armour scholar and especially for his practical knowledge of projectile weapons. He was entirely self-taught, and in his teens he spent a great deal of time studying in the major museums in Stockholm, especially the Artillerimuseum (now the Armémuseum), the Livrustkammare, the Nordic Museum and the National Museum. In this way he acquired not only much knowledge but also many contacts in the museum world who helped him in later life.

By the 1920s he was actively collecting and dealing in arms and armour and was frequently called in as an expert consultant by the major auction houses of Stockholm. From 1917 he worked as a volunteer for the Artillerimuseum, on a part-time basis, helping to document and arrange the inventories of the collection of arms and armour, but it was only in the early 1930s that he began to receive any payment for his work. In 1937 he was employed as a full-time curator of arms and armour, a position he held until his retirement in 1954. Alm began to write and publish on the subject of arms and armour in the 1920s. In 1927 a number of articles he had written for shooting magazines were collated into his first book *Vapnens Historia* (*A History of Weapons*), the first general survey of the subject to be published in Swedish. The book was criticised for lack of structure, which is hardly surprising given its origins as a series of articles, and for lacking source notes. Alm responded to these criticisms by producing some of his best and most important work in the following decade. These included: in 1932 *Blanka Vapen Och Skyddsvapen Från Och Med 1500-Talet Till Våra Dagar*) (*White Arms and Armour from and including the 16th Century to the Present Day*); and in 1933 and 1934 the two parts of *Eldhandvapen* (*Firearms*). Thereafter he concentrated upon writing articles, some in co-operation with other Scandinavian scholars, like Dr Ada Bruhn Hoffmeyer. These appeared in a number of journals, especially *Kulturhistoriskt lexikon för nordisk medeltid* (*Journal of the Cultural History of Medieval Scandinavia*) and the Journals of the Armémuseum, the Livrustkammare and the Vapenhistoriska Sällskapet (*the Swedish Arms and Armour Society*).

Josef Alm's contribution to the development of the study of arms and armour both in Scandinavia and around the world was very considerable. He introduced many in his nation to the subject, and both in what he wrote and in what he did in the Armémuseum he always tried to make the subject more popular and more accessible. He had a vast knowledge of projectile weapons, especially of the crossbow, and it may be that his most enduring and important achievement is and was the survey of European crossbows here translated into English.

Scandinavia

Alm's work on European crossbows inevitably concentrates upon what he knew best and the records which were most accessible to him. Thus his survey is weighted towards the Scandinavian countries and to Sweden, his native land, in particular. This brief resumé is intended to introduce the general reader to the often complex history of this area and thereby help to make Alm's text more immediately understandable. More detailed historical and geographical notes occur as necessary throughout the translated text.

The name *Scandinavia* first occurs in the writings of Pliny the Elder in about AD 75. To most people that name describes the area around the Baltic Sea which was the home of the Vikings. Today, Scandinavia consists of the four separate countries of Denmark, Finland, Norway and Sweden, the last three on the northern shores of the Baltic Sea, the first situated on a large tongue of land, and its associated islands, which thrusts into the Baltic from its southern shore.

In the period with which Alm concentrates, from the tenth to the seventeenth centuries, things were rather different. For much of this time Denmark was the dominant power, with its ruling house, whether by victory in battle or by intermarriage, sometimes controlling all and sometimes some of the other Scandinavian countries and also much of what today we think of as northern Germany. From 1157 the Valdemar dynasty gradually consolidated the power of Denmark and in 1397, by the Union of Kalmar, the kingdoms of Norway and Sweden were united with Denmark. The union of Norway and Denmark continued until 1814 (when Sweden took control of Norway). This is all the more remarkable since while Denmark, like Sweden, was an elective monarchy, Norway was a hereditary one. The union of Denmark and Sweden did not last nearly so long. A Swedish nationalist movement developed soon after the union and from 1434 there was a series of rebellions against Danish rule culminating in the establishment of lasting Swedish independence by Gustav Vasa in

1521. After establishing its independence, Sweden gradually took over as the dominant Scandinavian power and for much of the 17th century the Baltic was a Swedish lake, with Swedish territory nearly all the way around it.

It is also important to remember that until the state of Finland was created in 1917 there were only three not four Scandinavian countries. Until the mid-12th century Finland was unclaimed territory. Then it was taken by Erik IX of Sweden (1155-1161) and remained Swedish territory until 1714. From 1714 to 1721 and again from 1742 to 1743 Sweden lost control of Finland to Russia, and in 1808 it again passed to Russia, which remained the ruling power until Finland declared independence during the Russian Revolution in 1917.

However, if Sweden controlled Finland for most of the period about which Alm writes, it did not cover all the territory which today we think of as Sweden. Until the treaty of Roskilde in 1658 the southern and western seaboard of modern Sweden was not Swedish, apart from the estuary of the Göta Älv on which today stands Gothenburg. To the south of Gothenburg what are now the Swedish coastal regions of Skåne, Halland and Blekinge were part of Denmark, and the coastal area to the north of Gothenburg, now the region of Bohuslän, was part of Norway. In the north, far more so than today, the frontiers of Norway, Sweden (including Finland) and Russia were ill defined, the area being sparsely populated by a small population of migrant Lapps. To the south of this area most of Sweden was also very thinly populated with large tracts of forest separating the main areas or population in the north and centre of the country. Roads were few and dangerous and therefore the sea, lakes and rivers were much used for transport. This explains why many of the castles mentioned by Alm command waterways, which were then the main arteries of communication.

G M Wilson
Master of the Armouries

European Crossbows: A Survey by Josef Alm
1 Origin and primitive European types

During the Middle Ages, the bow, which consisted of a strung wooden stave, was known as the handbow in both Sweden (109:190) and Norway (62:135, 139, 140). It dates back to the later Stone Age, and was for a long time the only hand-held projectile weapon. A more highly developed type of bow is represented by the crossbow, which is a bow equipped with a wooden stock or tiller. In the primitive crossbow the top of the tiller has a hollowed-out recess or notch for the drawn string. Presumably, to begin with, the string was slipped out of this notch by hand, but even some of the most primitive of crossbows had a simple lock to release the drawn string.

A bow must be drawn immediately before the shot, whereas a crossbow can be kept spanned for a considerable period. This is frequently a great advantage, particularly when hunting. It is easier to aim with a crossbow than with a bow. To draw a bow, the string is pulled back with only one hand, and the force of the shot therefore depends on the strength of the shooter. However, even on primitive crossbows the string can be drawn into its spanned position with both hands. Because of this the crossbow can be more powerful than the bow. Its power can be increased still further if a special spanning mechanism is used. A heavy arrow for a bow, intended for use in battle, weighed only a little less than a crossbow bolt (63:115). Nevertheless, since the crossbow had a stronger bow, the effect of a crossbow shot was frequently appreciably greater than the effect of a bow shot over the same distance. On the other hand, a bow could be shot far more rapidly than could a crossbow.

It appears that the European crossbow evolved from the catapults intended for shooting arrows used in ancient Greece and Rome. The lighter types of these machines were very similar to crossbows (56:115-120) and it was a very short step to make such a machine in a portable form. However, it is also possible that the crossbow evolved from certain types of animal traps.

In other parts of the world, the crossbow has been used since very ancient times. Indeed, it is still used in parts of China, and in large areas of eastern Asia, from the Chukchi region in the north to the Nicobar region in the south.[1] In these regions the crossbow has appeared in many forms, from very primitive types in the Nicobars (51:181) to extremely ingenious repeating weapons in China (51:173f). The crossbow can also be found in certain parts of western Africa, although it only reached these areas at a comparatively late date.

Crossbows, apparently for hunting, are depicted on reliefs in France which date from the 4th century AD, and it may therefore be assumed that the European crossbow originated in what today we know as France. However, true military crossbows or arbalests (*arcubalistae*) were certainly used in the Roman army at this time as well (19:401-402). During the Migration Period such weapons are also believed to have been used by the Goths (56:423).

However, after about AD 500 the crossbow seems to have gone out of use in Europe, or at least to have been used very little. It is only after AD 900 that it once more appears, or is named specifically. This re-appearance seems to have

---

1 Chukotskiy Poluostrov is a region at the north east corner of Asia, bordering the Bering Straits. The Chukchi Sea begins to the north west of the Bering Straits. The Nicobar Islands lie betwen the Andoman Islands and the north of Sumatra in the Bay of Bengal.

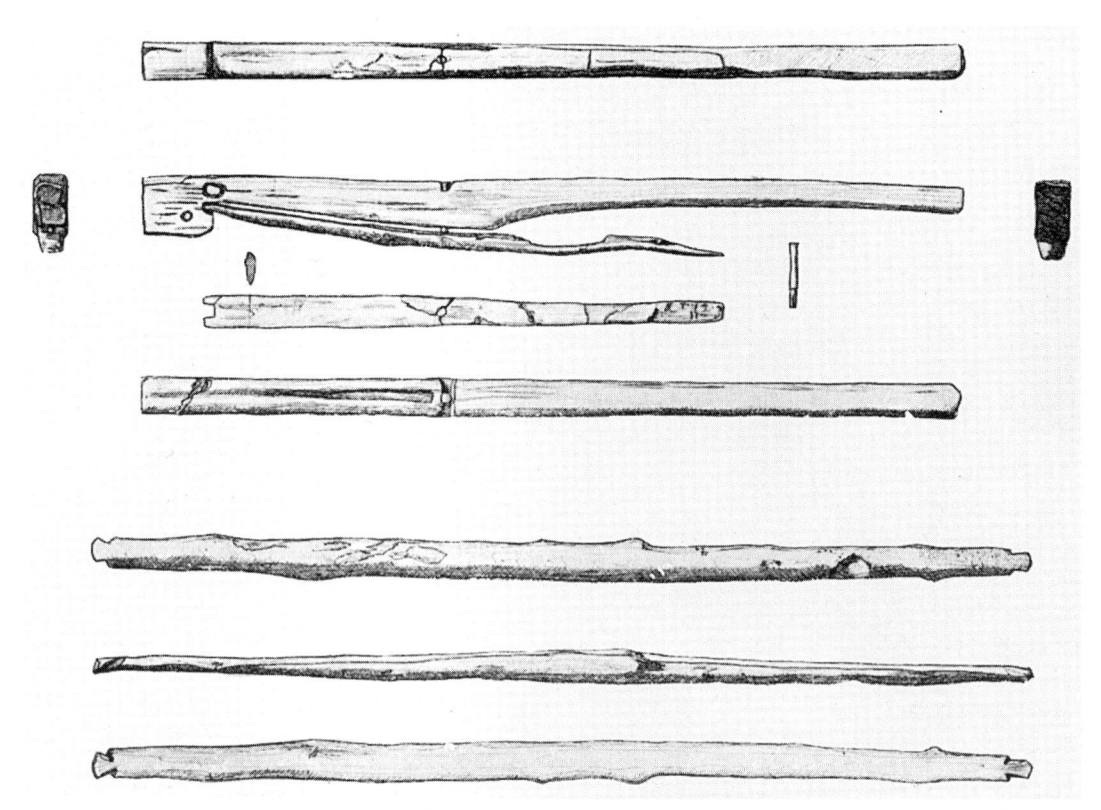

Figure 1 Crossbow found in the moat at Lillöhus in 1941. Perhaps from the peasants' revolt of 1525. Kristianstads Länsmuseum [Provincial Museum of North Skåne], Kristianstad.

taken place in western Europe where, for example, such weapons were used in the siege of Senlis in AD 947, and at Verdun in AD 985. (75:44).

The [Swedish] word for crossbow, *armborst*, comes from the Latin term *arcubalista*, which was used to define a projectile-engine fitted with a bow as distinct from those engines in which the propellant force was derived from tightly twisted ropes that acted upon levers. The *arbalestre* (79:659) or *arbaleste* (79:658) of medieval French is closely related to the terms *arbörst*, *arburst*, *arbyrst*, and *arborst*, (whether written with or without an *h* at the end), that were in common use in Sweden during the Middle Ages (10-12, 57) and in the 16th century (37-46). The terms seem to have come to Sweden via northern Germany, where the form *arborst* is recorded (18:74-75). The modern Swedish name *armborst* comes from Low German (18:74, 76). In medieval High German the crossbow was styled *Armbrost* or *Armbrust*. In medieval Latin the crossbow is called *balista* or *ballista*, terms which we shall discuss later in greater detail.

During the First Crusade crossbows and crossbowmen are reported to have made an important contribution to the Siege of Jerusalem in AD 1099 (56:116). At this time crossbows appear to have been rather primitive weapons with bows of wood. The Byzantine princess Anna Comnena gives the following description of the crossbow used by the Crusaders in the last decade of the 11th century:

> '*The* tzagran (or tzangran) *is a bow used by the barbarians that is absolutely unknown to the Greeks.*[2] *This long-range*

---

2 Anna Comnena (born 1183), daughter of the Byzantine emperor Alexius I (ruled 1081-1118), wrote 'an account of my father's deeds which do not deserve to be consigned to forgetfulness'.

7

projectile weapon is spanned by the user lying down on his back and putting one foot on each of the bow's arms and then drawing the string toward him with both hands. Along the middle of the tiller a semicircular groove is gouged out, running from the string (in its spanned position) to the centre of the bow, and along this groove the arrows are shot. The arrows which are usually used with the tzagran *are very short, but thick and with a heavy iron head. When the string is released it drives out the bolts with such great force and speed that they do not rebound from any object they hit. They have been known to pierce a shield, cut through a heavy iron breastplate and continue their flight on the other side, so great is the force with which these bolts are shot. Indeed, arrows of this type have been driven right through bronze sculptures and, when they have hit a city wall, their heads have either come out on the far side or have remained buried in the stones of the wall. The* tzagran *is thus an invention of the devil. The unfortunate man who is hit by it dies without feeling the mighty blow'* (9:68-69, 83:2)

Anna Comnena considerably exaggerated the effects of these bolts. To span a crossbow in the manner she described finger-stalls were frequently used in Europe to protect the inner sides of the fingers (75:60,61). The same spanning method 'with hands and feet' is still used in Siam [Thailand], (29:91), Cambodia (120:428), in other parts of south-east Asia, and also by various of the hunter peoples in Africa. In Africa the shooter sits down on the ground with both feet on the bow, one on either side of the tiller, and pulls the string back with both hands (27:279). In so doing he pivots his torso markedly in a backward direction, which gives us a reason to agree with Anna Comnena that the shooter lies on his back while spanning. A very strong crossbow can be spanned in this simple fashion.

It is difficult to say when the crossbow first came to Scandinavia, but those who lived in these northern lands were not slow to furnish themselves with novelties where arms and armour were concerned. In the *Jomsvikingasaga* it is related that bows and '*lock-bows*' were used in the battle at Hjörungsvåg in 986 (54:114).[3] In the 12th and 13th centuries in Norway '*lock-bow*' was a common name for the crossbow (62:135, 139-140).

The most primitive of all European crossbows yet discovered come from Skåne and from Norway, and were made entirely of wood.[4] Examples from Skåne (figure 1) were found in the moat at Lillöhus near Kristianstad.[5] They had very powerful bows. In the one illustrated, the bow is about 92cm long, 4cm broad, and 3.5cm thick at the middle. The tiller is about 81cm long. The groove for the bolt runs from the lock to the fore-end. The tiller has a spanning-notch on its upper surface about 29.5cm from the fore-end . Unlike other crossbows the tiller has a perfectly flat fore-end. Presumably the belly of the bow must have had a flattened middle, so that it could be placed directly against the fore-end of the tiller. It would have been held fast by lashings which went through the large transverse hole through

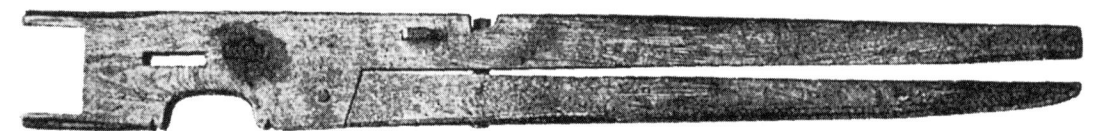

Figure 2 Tiller of a Norwegian 'whale-bow'. Nordiska Museet [Nordic Museum], Stockholm.

---

3 Hjörungsvåg, now Liavåg, is part of Storfjorden Möre fylke, Norway.
4 Skane (Scania) is the most southerly region of Sweden.
5 Lillöhus castle was built in the 1480s and was demolished in about 1650.

the tiller just forward of the front of the trigger. The purpose of the lower and more forward hole through the tiller is unknown and does not occur on the other tiller.

The lock consists of a long lever-trigger made of wood, and of a short wooden pin of circular cross-section which is mortised into the trigger at its lower end. There is a vertical hole in the tiller for this locking-pin, running up into the spanning-notch. Well forward of this the bottom of the tiller has a downward extension, the rear end of which has recesses cut into either side. The front end of the trigger is divided into two lips, which were obviously placed on either side of the recessed section of this heel. The trigger is pivoted to allow limited movement. Presumably the shape of both the front end of the trigger and of the recessed part of the near side of the heel kept the trigger from moving too far downward. It must be supposed that the locking-pin, which appears to be of greater diameter at its upper than at its lower end, prevented the trigger from slipping backwards so much that the lips at its front end could come out of their proper position. It is possible that a cord sling passed through the large hole towards the fore-end of the tiller and around the front of the trigger, as on the first type of west African crossbow which is shown in figure 6.

The locking-pin is long enough to allow the spanned string to press it down below the underside of the tiller and for the trigger's rear end to be moved a little away from the tiller through this action. When shooting, the tiller and trigger are held in the right hand and the trigger pressed upward against the tiller. This causes the locking-pin to push the string out of the spanning-notch and so shoot the bolt. For details of this crossbow I am indebted to

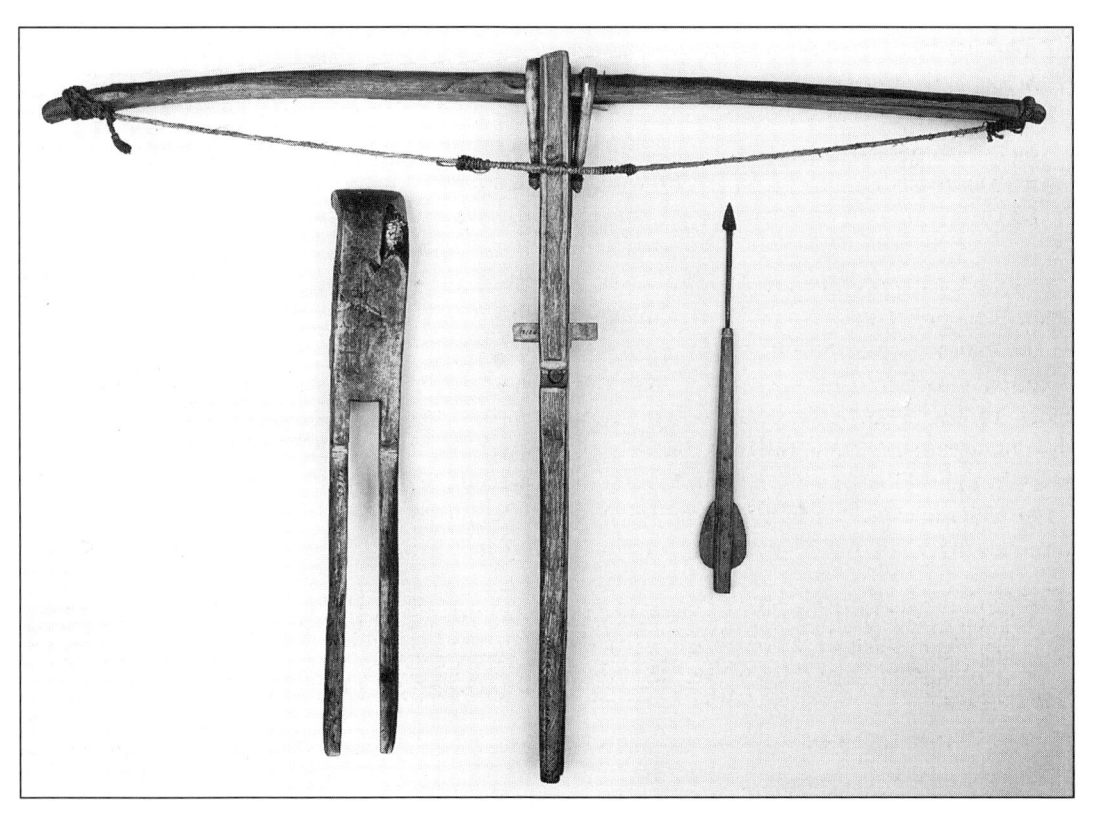

Figure 3 Norwegian 'whale-bow' with spanner and bolt. Spanner dated 1885. Nordiska Museet [Nordic Museum], Stockholm.

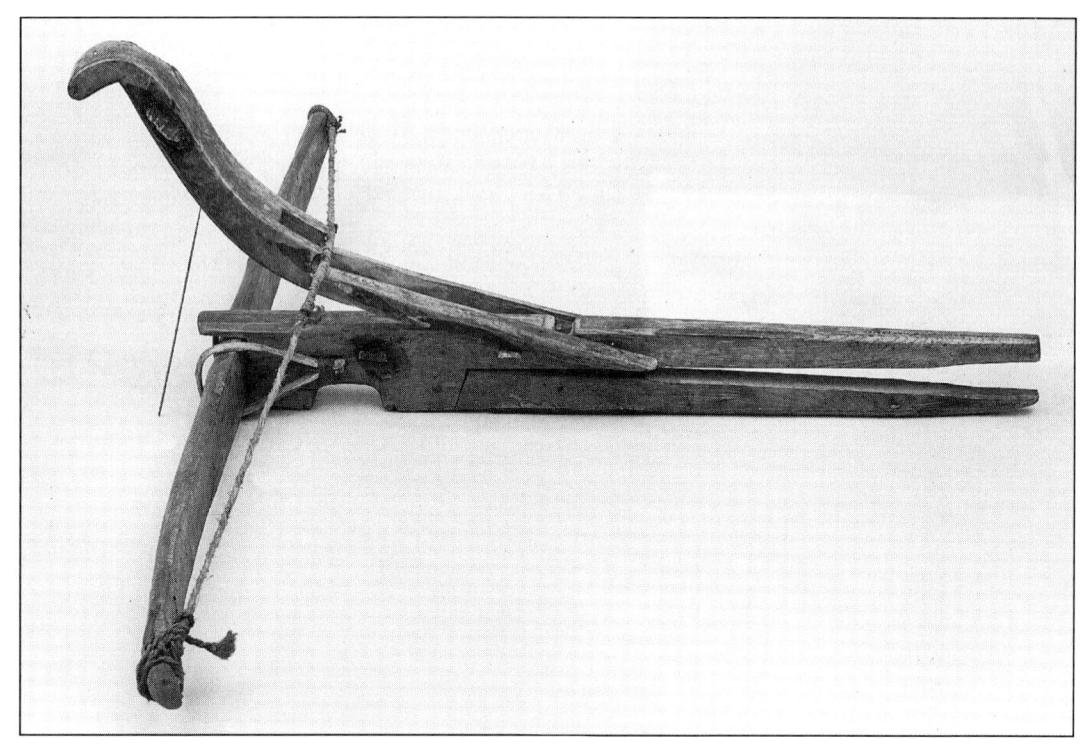

Figure 4 How a Norwegian 'whale-bow' is spanned. Nordiska Museet [Nordic Museum], Stockholm.

Thorsten Andersson, Curator of the museum at Kristianstad.

As well as the crossbow described above, a crossbow bolt was also found at Lillöhus. This has a short-socketed head with a tip of rhomboidal cross-section, and a shaft with flights set at an angle. Bolts of this type were used from the 14th to the 16th centuries. However, although the crossbow and bolt from Lillöhus must date from the late Middle Ages, the form of the crossbow is so primitive that it may be representative of those made at the beginning of the Middle Ages. Presumably these simple crossbows continued to be used for a very long time by poorer farmers for hunting, warfare, and revolt. They were probably spanned by using both hands and feet.

The Norwegian whale-bows (figures 2 and 3) prove that the Lillöhus crossbows were practical weapons for both hunting and warfare. These whale-bows were used for whaling in the Bergen area until about 1900. Whaling with bows and the bolts used for this purpose are described in Chapter 7.

The Norwegian whale-bows were generally constructed in the same way as the crossbows from Skåne just described. The only difference is that the long lever-trigger is fastened to the tiller at its front end by means of a wooden peg. The tiller of the example illustrated as figure 2 is of ash and has a bolt-groove from the lock to the fore-end which has a seat for the bow. Certain details, like the securing of the bow, and, particularly the spanning mechanism, are clearly borrowed from later crossbow designs.

The bow is made of yew and is attached to the tiller by a pair of stirrup-shaped wicker ties of osier or hazel, which are bound together at their ends. One tie is attached to either side of the stock. The ties are held fast at their rear ends by a wooden wedge that runs through a transverse rectangular hole in the tiller a short distance from the rear of the bow. These are clearly copies of the iron mountings, known as

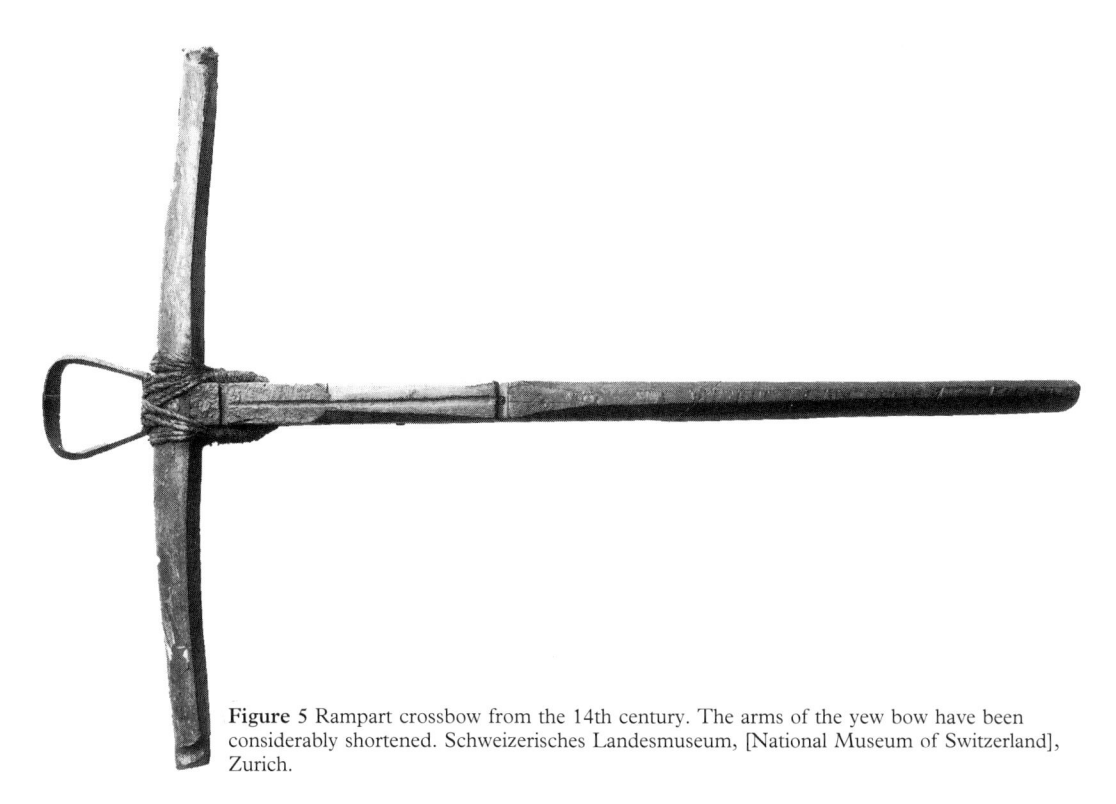

Figure 5 Rampart crossbow from the 14th century. The arms of the yew bow have been considerably shortened. Schweizerisches Landesmuseum, [National Museum of Switzerland], Zurich.

bow-irons, which secure the steel bow to the tillers of late medieval western-European crossbows. This sort or crossbow was predominant in Norway from the later part of the Middle Ages. The strings of whale-bows were most often made of fibres of fine hemp which were bound at the centre and served with yarn at both ends.

The whale-bow illustrated in figure 3 has a tiller 91.5cm long, and a bolt-groove 38.1cm long. The bow is 127cm long, 7.3cm broad at the middle, and 3.5cm thick. It has a flat belly, a rounded back, and narrow, flat edges.

The bow is spanned by a wooden device (figure 3) which is a primitive or simplified type of the goat's-foot lever which was in common use from the 14th to the 16th centuries. Some distance in front of the lock a hole runs transversely through the tiller in which is fitted a strong pin, which projects a good deal on either side of the tiller. In spanning, the crossbow is set with its butt against the ground, and is drawn as shown in figure 4.

It is difficult to determine whether these whale-bows represent an original or a degenerate form. Clearly they are a very old type that has continued to be used for a particular, specialised purpose, and which has consequently retained its primitive form. The goat's-foot lever and its transverse pin, however, must be later additions.

Related types of crossbows have been found in Central Europe and have been given early dates. Figure 5 illustrates a very large crossbow in the Schweizerisches Landesmuseum, Zürich, that has been dated to the 14th century. It has a wooden bow and a lock mechanism in which a wooden lever-trigger, now missing, moved a peg that pushed the string out of the spanning notch, in exactly the same way as on the two bows described.

The African crossbows described below suggest that similar crossbows must also have existed in southern and western Europe. The African ones are much like those from Skåne and Norway. They were, and indeed still are,

Figure 6 Tiller of a crossbow of the Niger type. After Balfour.

used in certain parts of western Africa. The Portuguese arrived there in the 15th century, and in the following century both the Dutch and English began to trade in these areas. On the Slave Coast, west of the mouth of the river Niger, was a kingdom, with Benin as its capital, which in the later Middle Ages was both large and powerful. From an early date European influence in Benin was strong. A large number of small statues, plaques of bronze with relief ornament, and ivory sculptures, have survived from this period. The decoration on these objects represents both natives, European warriors and merchants from the 15th century onwards. Many of the Europeans depicted carry muskets, but some have crossbows. If one may judge from these artefacts, the Europeans would appear to have brought with them crossbows of a type which greatly resembled the Niger types (15:347, 348, 353, fig.7). The natives in Benin, and presumably elsewhere on the coast, soon began to copy the European crossbows and to alter the design to suit their own ideas.

Two types of African crossbow are found today. The first comes primarily from the Benin area and from the district around the Niger delta. It is therefore called the Niger type (figure 6). In form this is very similar to the Skåne crossbow, but it differs from the Scandinavian type in two respects: the bow is fitted into a rectangular hole towards the fore-end of the tiller, where it is held fast by wooden wedges; and the trigger, made of a separate piece of wood as in Europe, is pivoted to the underside of the tiller by means of a loop of cord which runs through a hole in the tiller and around the front of the trigger. These crossbows have a bolt-groove in the tiller from the lock to the fore-end.

One typical crossbow of this type has a tiller 61cm long. Its bow is 68cm long with a rounded back, flat belly, and flat upper and lower edges. At the middle the bow is 4.46cm broad. One detached crossbow tiller of this type was found to be 85.09cm long (15:342-343).

The other form of African crossbow (figure 7) is known as the Fang type, taking its name from the Fan or Fang people amongst whom its use was first encountered. This type has evolved from that just described, but it does not constitute any improvement, rather the reverse. It appears in several variants. Some have a bolt-groove, approximately 7-8cm long, beginning at the lock, which becomes gradually shallower as it extends forward. Others have a short groove or recess for the bolt at the fore-end of the tiller. The trigger is not made of a separate piece of wood, but instead the tiller is split to the rear of the lock, the tapering lower limb serving as the trigger. Sometimes the two halves of the tiller are held together at the butt, either by means of an iron band or the like, but sometimes this is unnecessary because the tiller has not been split all the way back (figure 7).

In some cases the tiller is so shaped that it looks as though the lower limb of the tiller, which serves as a trigger, is made of a separate piece. Sometimes the tiller has a circle of ornament at the point where the Niger type has a hole for the cord loop that forms the trigger's hinge. In other respects the lock mechanism is identical to that of the Scandinavian types (15:339f).

Upon examination, a typical crossbow of this type proved to have a tiller 133cm long and a bow 78 cm. long. The bow was of rectangular cross-section, 3cm broad and 2.5cm thick at the middle. The string was apparently made of plant fibres and was reinforced at the middle

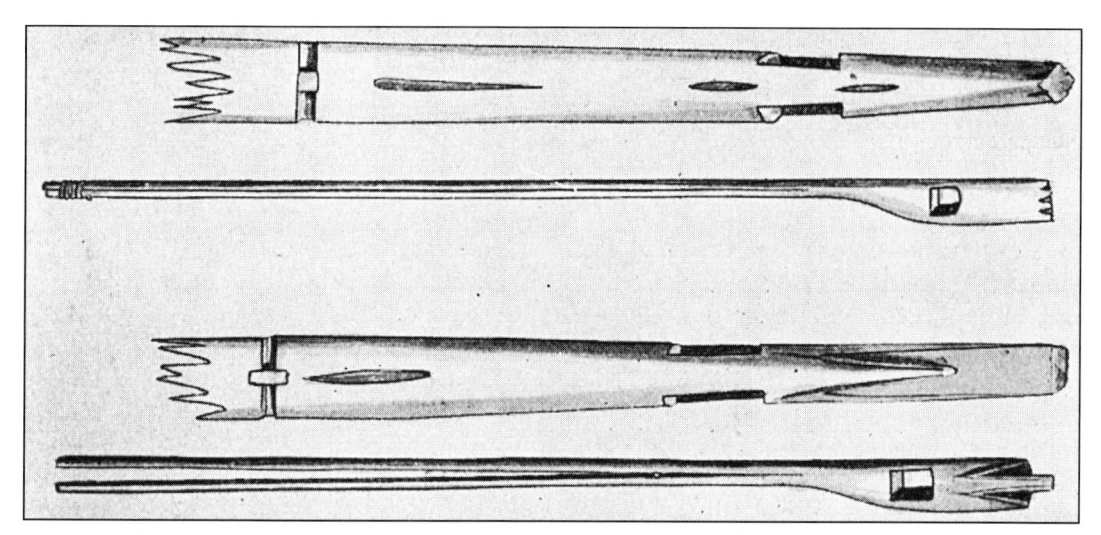

Figure 7 Tillers of crossbows of the Fang type. The forward portions of the tillers seen from above, and the entire tillers seen in protile. Folkens Museum Etnografiska [National Museum of Ethnography], Stockholm.

with yarn served on.

These crossbows often had powerful bows. In spanning, a piece of wood was set between the two rear limbs of the tiller to keep the trigger-peg out of the spanning-notch (15:321). The method of spanning varied among the different tribes, and presumably also depended on the power of the bow. Among the Fang people such bows were spanned, as noted above, by using both hands and feet (27:279). Other tribes appear to have had weaker bows. Among certain Pangwe groups, the bow was spanned by setting the butt of the tiller under the right arm, putting the knee against the bow, and drawing the string with both hands (110:140). At one time this same spanning method was probably also used in Europe.

The African crossbows were and are still used exclusively for hunting. In the 1860s such crossbows were occasionally used for hunting big game with iron-headed bolts about 60cm long. However, even then it was more usual to shoot light poisoned bolts made entirely of hardwood, and since then these have become the only type of arrow to be used. To prevent these light projectiles from moving while taking aim, a little wax or gum was smeared in the bolt-groove and the bolt was stuck to this (15:341). These bolts were fletched either with leather or some leaf material. They were used for hunting birds, squirrels, monkeys and other small animals (68:104). In Africa they did not learn that the crossbow is best suited to shooting very heavy projectiles.

2 The 12th and 13th centuries

The design of the crossbow was improved considerably during the 12th and 13th centuries. Most of the improvements seem to have been made towards the end of the 12th century, but did not come into general use until well into the following century.

The crossbow lock comprising a nut and a long trigger-lever (figure 8) apparently appeared at the end of the 12th century (63:115). The release nut is a cylinder, about 3cm long and 5cm in diameter. On the upper side is a lateral groove to take the string, and at right angles to this a slot which corresponds in width to the thickness of the end of the bolt. The top of the nut, therefore, appears to be two 'fingers', the slot between the fingers constituting a continuation of the bolt-groove. On the underside of the nut, opposite the groove for the string, is a notch for the front end of the trigger which is usually reinforced by a small piece of steel set into the surface. This nut fits exactly into a specially-shaped seat in the tiller. This seat has to bear the full pressure exerted by the spanned string. To begin with it was simply cut out of the wood of the tiller, but later its forward side was reinforced with horn. Sometimes it was made entirely of horn, often in two pieces, one at the back, one at the front.

Throughout the Middle Ages the nut lacked a spindle and was 'free-floating', held in place simply because it fitted into its seat. The long trigger-lever or iron was bent in two angles, its front end serving as a sear. Only towards the end of the Middle Ages was the trigger fitted

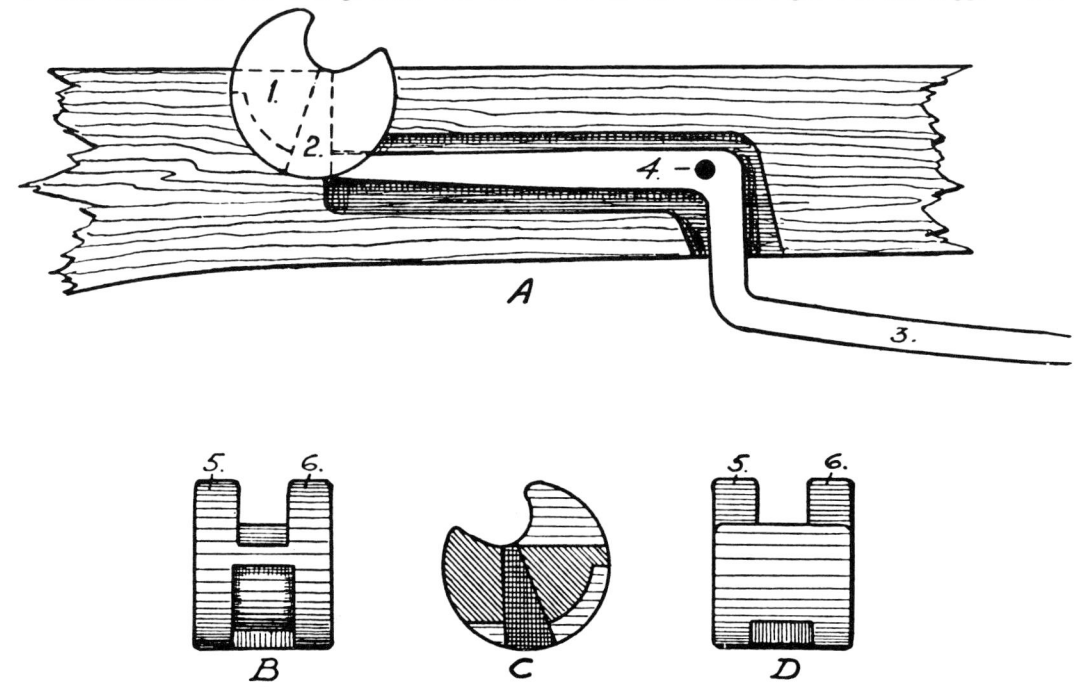

Figure 8 Crossbow lock. *A* - the complete lock; *B* - the nut seen from forward and *D* - seen from behind; *C* - the nut in cross-section. *1* - the nut; *2* - a steel wedge which forms the forward end of the sear notch; *3* - the trigger-lever; *4* - the trigger spindle or pin; *5* [& *6*] - the fingers of the nut.

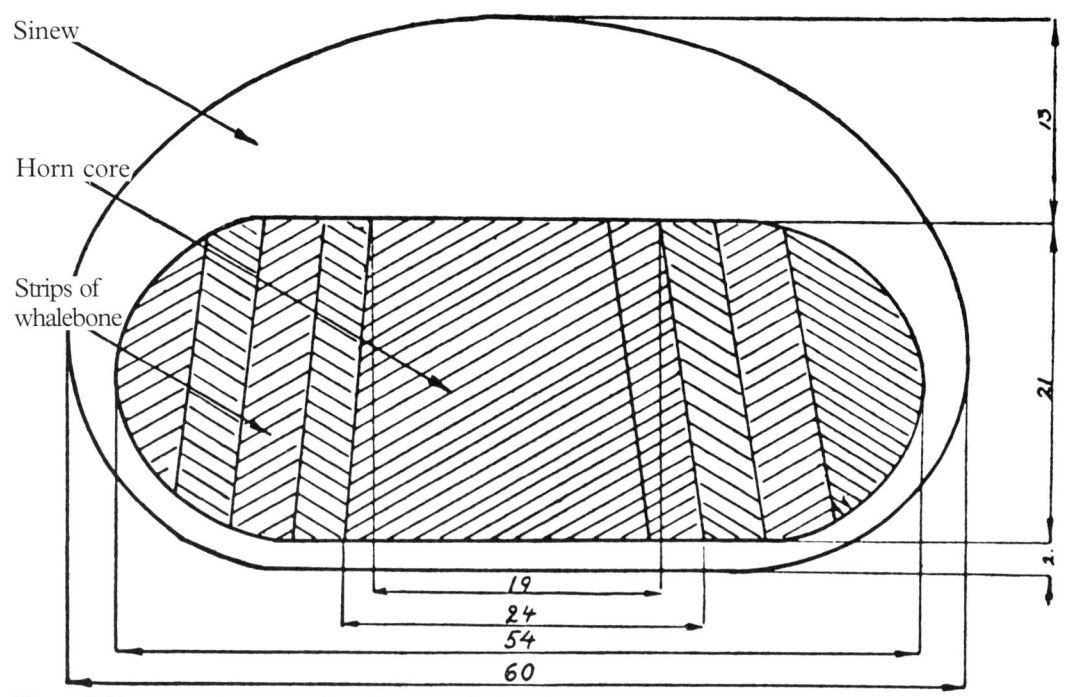

Figure 9 Diagrammatic cross-section of a composite bow for a crossbow, type 1. After Rohde. [All measurements are in millimetres].

with a flat spring that held its front end against the nut. On either side of the lock the tiller was reinforced with splints of horn, bone or iron. To begin with these splints were quite small, and indeed the oldest crossbows appear to have had no reinforcement whatsoever. Certain Roman projectile-engines had had similar locks (75:301).

The stirrup-like foot-loop (figures 20-21) also probably appeared at the end of the 12th century (63:115). Its back lay against the middle of the back of the bow, and was held fast to the bow with lashings.

To be very powerful a wooden bow had to be made both large and clumsy. Therefore, in the latter half of the 12th century the better crossbows generally began to be equipped with 'horn bows'. Such bows could be made considerably smaller and hence lighter than a wooden bow of the same power.

The 'horn bows' used in these European crossbows were copied from the extremely ancient Asiatic composite bows, the finest of all bows.

The centre of a bow is called the grip, and from either end of this extend the two arms. The ends of the arms are called 'ears' on composite bows, and in each of these is cut a nock to take the string. The side of the bow which is turned forward when shooting is called the back, and the side that faces toward the shooter is called the belly.

In a composite bow of Asiatic type the details of design and construction vary depending on the date of manufacture and the country of origin. Generally, however, such bows have a core of wood, quite thick in the centre, but decidedly broad and flat in the arms. To the belly of this core was glued a layer, made of strips or ribbons or horn. To the back of the core was attached a thick layer of sinews bonded together with a little fish-glue (13:82f).

Normally the composite bows of European crossbows lack the horn layer on the belly, which is usually found on Asiatic composite bows, this layer generally being replaced by horn or whalebone in the usually very thick

Figure 10 Fragment of a horn bow for a crossbow, type 2. After Rohde.

core, the composition of which varies, and is often complicated.

The German scholar Rohde described three different types of composite bows. The core is always surrounded by a layer of sinews that makes up from a third to a half of the volume of the bow (82:54f).

In the first type of composite bow (figure 9) the core consists of a long piece of horn with a trapezoidal cross-section, and a number of shorter strips of whalebone placed on either side of the horn and fastened to it with fish-glue. As is usual with composite bows, the glued surfaces are cut with grooves or channels that fit into each other, and which were presumably made using a plane with a saw-tooth blade. In this way the various parts are prevented from slipping against each other as the bow is spanned. The belly and back of the core also have grooves of this sort.

Such grooves also occur in the composite Asiatic bows. The core is enclosed on the back, sides and belly by a layer of sinews. In the centre of the back the sinew layer is 1.3cm thick, but it becomes thinner toward the ends and at the sides. The fibres of the sinew layer run longitudinally along the bow on the back and sides, but transversely on the belly, where the sinew layer is a mere 0.2cm thick. The horn core is somewhat longer than the sinew layer and the strips of whalebone. Most unusually, in the example described by Rohde, nocks are cut into both of the projecting ends of the horn section. The bow is clad with birch-bark, which was originally painted. Since no string is attached, the bow curves forward a little as usual. Rohde dates this bow to the 14th or 15th centuries (82:54), but it could be as much as a century older than this, especially as it is very much like a composite hand-bow.

The second type of composite bow (figure 10) was built up in much the same fashion as the first. However, in this type the core does not consist of one long piece of horn, but of delicate strips or slices of horn or whalebone glued together. This core is surrounded by a layer of sinews, into which the nocks are moulded by pressure (82:55).

In the third type (figure 11) the bow is composed of a great number of strips of horn or whalebone of different lengths which are glued together. As usual, the surfaces to be glued are cut with grooves. The core is constructed not in one but in two sections, separated at the centre by a section 6cm long, 3cm broad, and 1.2cm thick, made up of a number of pieces of horn glued together. Glued to both the upper and the lower sides of the core is a strip of yew-wood. This is surrounded, as usual, by sinew and birch-bark (82:55).

The English scholar Payne-Gallwey describes a similar type with the core in a single

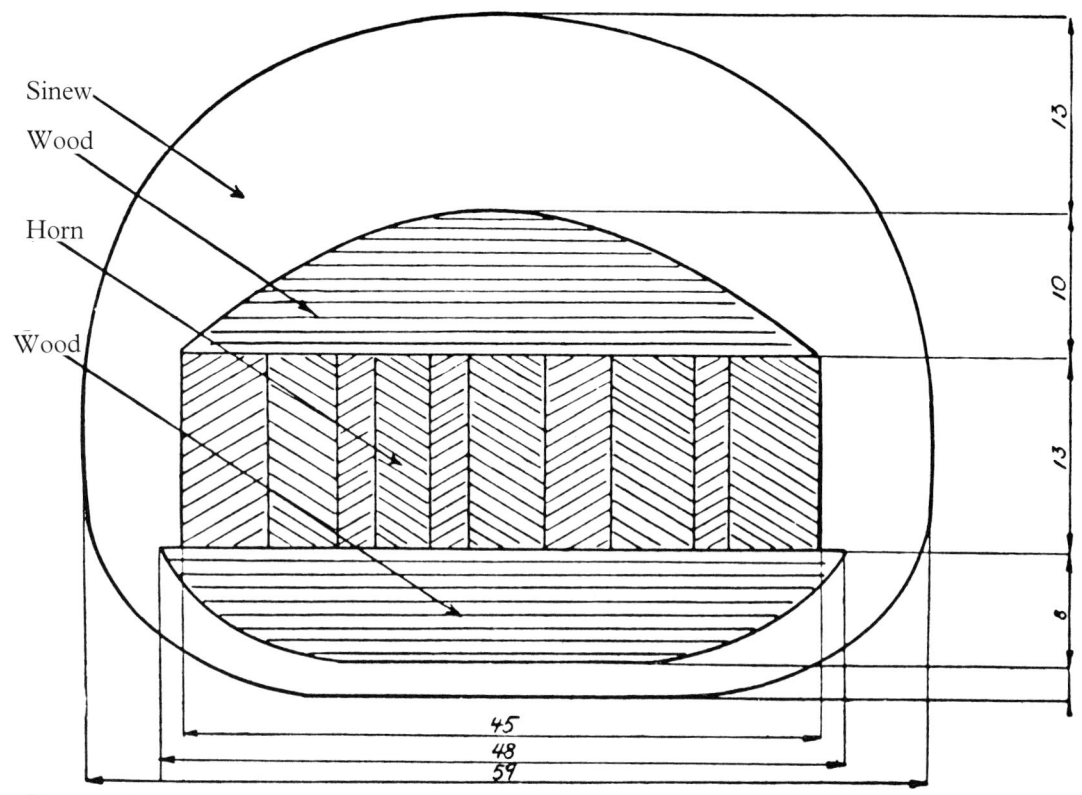

Figure 11 Diagrammatic cross-section of a composite bow for a crossbow, type 3. After Rohde. [All measurements are in millimetres].

section and glued together with about 20 pieces of sinew or strips of whalebone. As in the type described above, a strip of yew-wood is applied to the upper and lower sides of the core, and the whole is coated with sinews and finally with birch-bark (75:64).

A very different design appears to have been usual in southern Germany and Austria. One of these bows, which dates from the second half of the 15th century (figure 12), has, towards the belly, a core of young oak, about 6cm wide, and 1cm thick, which becomes narrower and thinner toward the ends. On the back the core is cut with grooves, and glued edge on to this side are 12 horn strips. These horn strips are all of different thicknesses and cross-sections, and are grooved on all sides. To these are attached a layer of sinew, consisting of a number of strips glued together. The belly side is clad with thick, untanned pigskin. The whole is overlaid with birch-bark to which is glued an outer coating of thick paper decorated with stamped designs (115:54).

The construction of composite bows for crossbows varied a great deal both during the Middle Ages and the 16th century, presumably depending either on the place from which the crossbow-maker derived his models or on his experience, judgement or taste.

Sinews from the necks of horses or oxen were used to make the sinew layers of composite bows (75:64, note 1).[6] In the lands of the Teutonic Knights principally goats' horns but also ox horns were used in the 14th century for making the cores of composite bows (79:655). The horns were first softened by being immersed in boiling water, then cut and pressed flat while soft, and finally sawn into slices or

---

6 Modern research and analysis suggests that leg sinew was probably more frequently used.

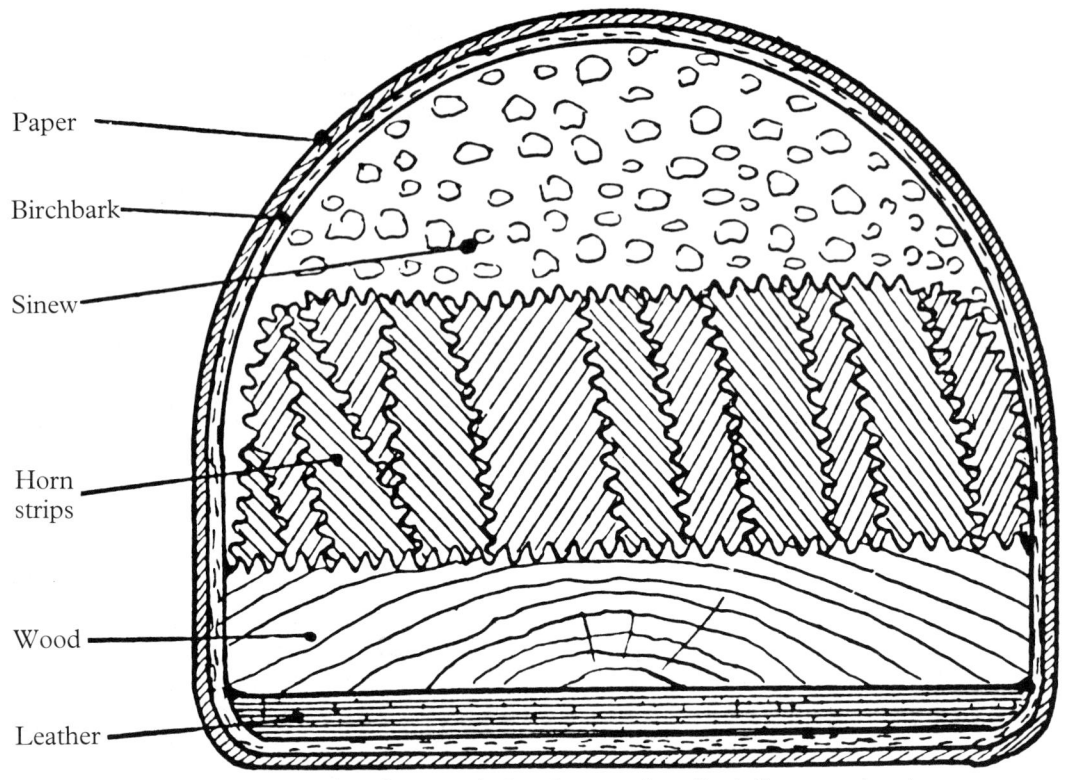

Paper
Birchbark
Sinew
Horn strips
Wood
Leather

Figure 12 Diagrammatic cross-section of a composite bow for a crossbow. South German or Austrian type. After Wegeli.

strips. In the inventories of the *Schnitzhäuser* or cutting workshops that existed in the areas controlled by the Teutonic Knights horn-saws, horn-files, and horn-presses are listed (79:655).

The composite crossbows described by Rohde are very similar in design to the composite Persian bows (13:89f). It is difficult to say when and where composite bows first began to be used for crossbows. Most scholars consider that the idea was brought back by the crusaders from Palestine and Syria (75:62). The crossbow did indeed play a great part in the Crusades. Crossbows were unknown in the Orient before the Crusades. However, during the crusading period, crossbows seem to have been made by the Arabs and to have been used by them to some extent, especially in fortresses. Thus, for example, after Damietta had been taken by the crusaders in 1219 powerful crossbows with composite bows, as well as many other crossbows and simple bows, were found there (63:174).[7]

One of the first English crossbow-makers known by name was Peter the Saracen, who made crossbows for King John in 1205 (75:62). This might seem to reinforce the theory that the

---

[7] The most informative source about Saracen crossbows was not available to Alm when he wrote his Survey. W.F. Paterson noticed this omission when editing the text and wrote the following note on Saracen crossbows to help the reader of this translation:
'The most important source is from a manuscript in the Bodleian Library, Oxford (Huntington Ms 264), and the relevant section has been translated by the French scholar Claude Cahen ('Un Traité d'Armurerie Composé pour Saladin', Bulletin d'Etudes Orientales, Vol XII, 1947-8, pp103-163). The work by Mardi ibn 'Ali at-Tarsusi was dedicated to Saladin and would therefore have been written before the latter's death in 1190. From this we learn that the Saracens made crossbows with bows both of yew and olive, and of composite design that followed the normal Mideast pattern. They were made both with and without stirrups. The waistbelt, fitted with a two-hooked claw (khattaf) appears to have been the most common method of spanning. The lock mechanism employed the nut (jawzah).'

Figure 13 How the crossbow string was fitted. From a painting of the 15th century in the Escorial. After Rohde.

composite crossbow originated in the Near East. However, the Germans were already acquainted with composite bows. They had been used by the Huns and later on by the Avars, Magyars, Tartars, and Kumans. In addition, such bows were used in the Byzantine Empire and by the Moslems in Spain and Sicily. Crossbows probably began to be fitted with composite bows at about the same time in a number of different parts of Europe. However, crossbows of wood were used side by side with composite bows until much later. The former were quite naturally a great deal cheaper and easier to manufacture than the latter.

Composite and wooden bows were attached in the tiller by a lashing of sinews or cords which passed through a hole in the tiller a little to the rear of the bow. As we have already noted, the stirrup for the foot was also secured to the bow by means of lashings. In later times, at least, the lashing consisted of a long cord 4-5 mm. in diameter, which was basted together with ten to 12 waxed threads (82:56). The bow was not fitted vertically into the tiller, but at such an angle that the ends of the bow rose a little above the plane of the tiller's upper surface. This was so that the string would merely lightly touch the upper side of the tiller, but not rub against it. Crossbow strings were most frequently made of a single long yarn or cord of linen or hemp (75:110-113). Up to a rather late date in Germany strings were made of a fine, well-waxed cord at least 150m long. A pair of pins was driven into a floor 4-5m apart, and the cord was wound tightly between them. The hank produced in this way was folded several times until the string was the correct length. In this way the string came to consist of between 60 and 80 strands. Each of the string's loops was reinforced with a wrapping of thread and was then clad with thread that was four to five times as thick as the threads in the string itself. This cladding was fastened on the outside with special knots, so that specific patterns developed. At the centre, where the spanned string was to rest against the fingers of the nut, the string was tightly served with thread. Between this serving and the loops the string was sparsely bound together with the same sort of thread. All the threads were waxed (75:110-113, 82:56-57). Some Asiatic composite bows have strings of a very similar design (13:92).[8]

The ends of an unstrung composite bow usually point forward. In order to get the bow into a position in which it could be strung considerable force was required. For this a special, strong, auxiliary string was used, which was fastened to the bow a little inward of the stringing notches, figure 13. To start with, this auxiliary string was not drawn. The crossbow was then fastened to a sort of draw-bench, and the auxiliary string was drawn back by means of a hook, linked to a very powerful screw, which was turned by means of a couple of handspikes. When the bow had been drawn sufficiently, the regular string could be put on (82:57). A crossbow is always slightly drawn, even when it is not in its spanned position. This is particularly pronounced in the bows that bend forward when they are unstrung, so that when they are strung they may be regarded as being half-drawn.

In the 15th century the distance between the string and the rear of the fingers of the nut was most frequently around 15-20cm, but earlier it was evidently greater. Crossbows of the Middle Ages have stocks generally 80-90cm long.

As crossbows were made stronger it was no

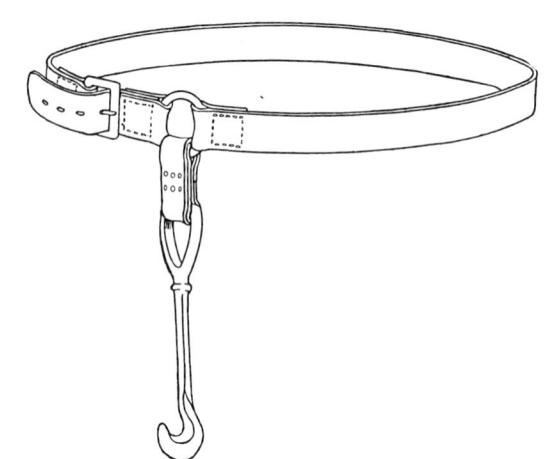

Figure 14 Spanning belt with spanning hook.

longer possible to span them by means of the hands alone. The spanning hook with one or two claws was then invented. It seems that to begin with the spanning hook had a grip at its upper end. In spanning, one placed a foot in the stirrup, hooked the claws of the spanning hook over the string, and, with both hands on the grip of the hook (63:115, note 1), drew it up to the spanned position. In this way one got a better hold, but one could scarcely have developed more power than with the hands alone.

Soon, apparently in the latter half of the 12th century, the spanning hook came to be used with a belt. The hook was attached to a short, strong strap, which was permanently fixed at its other end to a large iron ring that joined together the two elements of the spanning belt (figure 14). To span the crossbow, one set it down with the bolt-groove towards one's body, put a foot in the stirrup, either bent forward or

---

8 W.F. Paterson had great practical experience of making strings for bows and crossbows. While working on Alm's text he prepared the following note for readers:
'Though the string is an essential part of a crossbow, little has been written about its construction. Strings are of two main types: those of the heavy war and hunting crossbows; and those of the lighter examples, mainly used for hunting, which include the pellet and bullet-shooting crossbows. The problem with any string is to get the pins or pegs, around which the skein is wound, the correct distance apart. A regular maker or strings would, or course, have a simple rule by which to work, related to the width between the saddles of the nocks of the bow. An aspect that requires considerable care, and is difficult to achieve, is getting an even tension on all strands when the string is fitted on the bow. An additional problem is how the end loops were reinforced so that the thickness of the strands round the nocks was about the same as that of the main skein. Available evidence suggests that there were at least three different methods of achieving this, apart from that shown by Payne-Gallwey, who admitted that he did not know how it was done (75:10) Criticism must be made of Alm's figure that gives such strings 60-80 strands. About 200 is closer to the mark for the heavy strings. As regards his comment about the use of special knotting, this is nothing more than half-hitches put on, alternately, right- and left-handed to form what is known as cockscombing. It gives a better finish than a straight, simple serving around the curve of an end loop.'

went down on one knee, hooked the claws of the spanning hook over the string, and straightened or stood up, thus drawing the string into its spanned position using the full strength of the body.

That part of the nut behind the fingers was usually cut out, so that when the bow was spanned it stood a little above the upper surface of the tiller. When spanning the bow one rotated the nut so that the rear surfaces of the fingers lay below the top of the tiller. When the string was then drawn backward it pressed the rear of the nut backward and downward till the forward end of the trigger could enter the notch cut in the nut (75:127).

Once the bow had been spanned, one pressed the rear of the bolt between the 'fingers' of the nut so that the end rested against the spanned string. As a result, the part of the bolt-shaft to the rear of the flights usually had flattened sides.

When shooting one either put the butt of the crossbow tiller on the shoulder, or thrust it under the armpit. The head of the bolt was most frequently of rhomboidal cross-section, and was fixed so that, when the bolt lay in the groove, one of the angles of the tip lay upward and served as a sight. The crossbow was held with the left hand a little forward of the lock, with the right hand around tiller and trigger, so that the knuckle of the thumb was on top. The upper angle of the thumb served as a sight. Then one pressed the long trigger-lever upward against the tiller, whereupon the front end of the trigger released the nut, which, once freed was rotated by the string, which drove the bolt forward forcefully.

During the 12th and 13th centuries crossbows were very common in western and southern Europe, and soon in central Europe as well. In 1100 William II of England was killed by a crossbow bolt while hunting. His successor Henry I (1100-1135) had many crossbowmen in his service (75:45).

Under Louis VI (1081-1137) the use of the crossbow was widespread in France (79:643, note 1). Knights regarded the crossbow with scorn, but also with fear, for it was a weapon which made infantry dangerous.

The Church fully shared their view, and that of Anna Comnena, that the crossbow was an invention of the devil. The Lateran Council held in Rome in 1139 commanded that the crossbow should not be used against Christians, but only against heretics and pagans (63:111). This decree had little effect. To be sure, the Emperor Conrad III (1138-1152) ordered that the crossbow should not be used either by his army or in his domains (75:3), but other rulers, for example, the English King Stephen (1135-1154) and Henry II (1154-1189), had many crossbowmen in their service (75:45). Richard the Lionheart (1189-1199) was himself a great advocate of the crossbow and is supposed to have slain many enemies with the weapon (75:46). In the Third Crusade his crossbowmen fought against the Moslems in full conformity with the Church's bidding. However, thereafter he let his soldiers retain their crossbows unmolested in Europe, and he was himself killed by a crossbow bolt at the siege of the fortress of Châlus in France (75:3). Philip Augustus of France (1180-1223) employed crossbowmen both mounted and on foot (19:403). Even the popes were unable to ban the crossbow from their own armies (79:643/644, note 1). At the end of the 12th century Pope Innocent III repeated the ban on the use of crossbows against any other than heretics and pagans, but without the slightest effect (75:3).

Towards the end of the 12th century, as protection against the deadly crossbow bolts,[9] knights donned helmets with visors and a

---

9 Alm was here indulging in speculation. There is some evidence for the introduction of visors on some helmets in the late 12th century, and some isolated references to trunk defences which may be translated as "breast plates", but there is no evidence for the general introduction of breast plates until the late 13th century (see C. Blair, *European Armour*, London, 1958, pp.37-41. 47-8).

'breastplate' of iron, which was worn under or over a mail coat. At the same time horses, too, were equipped with armour, or *hästbrynjor* [literally 'horse-mail'] (63:35, 41; 62:139-140).

During the 13th century, in southern and western Europe, there is record of hand crossbows of two different types: *arbalista ad unum pedem* and *arbalista ad duos pedes*. The first type had a stirrup which was meant for only one foot and which I shall call a crossbow with a one-foot stirrup. Crossbows of the other type had stirrups so roomy that one could put both feet into them (75:60), and these I shall call crossbows with a two-foot stirrup. The crossbows with one-foot stirrups were smaller, shot lighter bolts, and could perhaps have been spanned with the hands alone, or possibly with a spanning hook. For crossbows with two-foot stirrups the spanning hook and spanning belt were used. These shot quite heavy bolts and appear to have been used in fortresses and fortified towns; they were equipped with either wooden bows (*balistae ligneae ad duos pedes*) or composite bows (*balistae de cornu ad duos pedes*) (48:326).

Naturally the crossbows with a one-foot stirrup could also have either wooden or composite bows. Sometimes there is mention of crossbows that could shoot two bolts at the same time (48:326). In old French such weapons were called *arbaleste gemelle* (79:659). Presumably they were usually large weapons. One can be sure that crossbows of older types continued in use, and that composite bows had by no means driven out wooden bows.

There is not much published information regarding crossbows in the 13th century. In 1239 the Emperor Frederick II ordered a sea-captain who was bound for Acre to buy there as many crossbows with two-foot stirrups (*balistas de duobus pedibus*) as he could lay hands on (63:175).[10]

In 1269 crossbows with two-foot stirrups (*balistas de duobus pedibus*) and crossbows with one foot stirrups (*balistas de uno pede*) are mentioned at Piacenza (63:175).[11] In 1224 the Genoese crossbowmen, so renowned during the Middle Ages, are said to have had crossbows with composite bows (63:113, note 2). In 1246, 500 Genoese crossbowmen were serving against the troops of Milan, and proved so dangerous to their enemies that in reprisal the Milanese put out one eye and cut off one hand of every crossbowman they captured (75:62).

In England King John (1199-1216) and King Henry III (1216-1272) used many crossbowmen, mostly foreign mercenaries. However, after the death of Henry III the crossbow began to be superseded by the famed longbow (75:46).

The crossbow was in general use in all the German cities during the 13th century, especially in the cities of the Netherlands and the Hanse. Shooting guilds with Saint Sebastian or Saint Maurice as their patrons were established at an early date (19:403). Shooting at the popinjay, which later came to be so highly prized, began in the thirteenth century. At Schweidnitz a shooting competition of this sort took place for the first time in 1286 (91:328).[12] As a result, the citizens of these cities were often good shots with the crossbow, a fact which their enemies frequently found to their cost.

In addition to hand crossbows there were also large, non-portable crossbows that were used to defend cities and fortifications, or to besiege them. Sometimes these were also used in battle. For shooting they were set upon stands, and sometimes they were moved about on carts. They were called windlass crossbows: *balistas de torno* and *arbalestes à tour*, both of which mean that they were spanned by means

---

10 Frederick II was elected Holy Roman Emperor in 1215. He was also King of Sicily. Acre is on the coast of Israel just north of Haifa.
11 The town and province of Piacenza lie just south of the River Po in northern Italy.
12 Schweidnitz is a town in the south west of modern Poland, now called Swidnica, which lies some 25 miles south west of Wroclaw.

of a windlass. It is difficult to say what sort of a windlass was used. Payne-Gallwey considered that they were spanned with a hook attached to a pulley, later called a *krihake* in Sweden. He seemed to believe that what were being referred to were hand crossbows (75:73, note 1).

In fact, however, it is quite clear from several sources that what is being referred to are large weapons, *magne arbaliste ad turno* (63:175, note 5). Probably the windlass was the same as what the Germans called *Haspelwinde*, a wooden cylinder which was rotated by means of transversely placed spokes. If this apparatus was mounted on a strong, short plank, there was no difficulty in equipping it with a strong rope sling or an iron socket, which could fit over the butt of the crossbow tiller to span the bow. The spanning was done by means of a rope which had a hook at one end, and was permanently fastened to the cylinder at the other. To span the bow, the hook was attached to the bowstring and the cylinder was rotated, whereupon the rope was wound onto the cylinder thus spanning the bow. The tackle was probably reinforced at quite an early date by introducing pulleys, but nothing is known about this.

In 1239 the Emperor Frederick II ordered a sea captain to buy as many windlass crossbows (*balistas de torno*) as he could at Acre. *Balistas de turno* are also mentioned at Piacenza in 1269 (63:175).

During the 13th century crossbow bolts seem generally to have had short, heavy heads of rhomboidal section, which were attached to the shaft by a socket. Very large bolts were used with windlass crossbows.

Like Richard the Lion-Heart, princes and distinguished soldiers in Denmark and Norway often carried crossbows during the 12th and 13th centuries. Thus, for example, Saxo says that Esbjörn Snare used such a weapon in a battle with the Estonians and Courlanders on Öland in 1170.[13] He shot three bolts at the enemy, but, as none of them hit the target he smashed the bow in a rage (87:152). Perhaps his weapon was not well designed. Saxo also reports that another Danish chieftain, Sune Ebbesön, used a crossbow during the same year against the Wends at Julin (87:161).[14] The Church did allow the use of such weapons against pagans, which is what the Estonians, Courlanders, and Wends were at this time.

In 1231 Valdemar, king-elect of Denmark, was accidentally shot dead by a crossbow bolt (18:74). In the Jutland Law adopted in 1241 it was laid down that the helmsman on every naval vessel should have, in addition to other weapons and armaments, a crossbow with three dozen bolts *'and a man who can shoot with it if he cannot do so himself'* (55:361).[15] This passage appears to indicate that at this time the crossbow was still something of a novelty to the Danes.

It is recorded that in 1296 citizens in revolt at Copenhagen shot sharp bolts into Copenhagen Castle, where *'their spiritual and temporal ruler'* Bishop Jens Krag of Roskilde, was staying (18:77).

At some time between 1100 and 1123 the Norwegian King Sigurd Jorsalafar said scornfully to his brother Östen, *'You could not span my bow, even were you to brace yourself against it with both feet'*.[16] It seems, therefore, that Sigurd's bow was a crossbow of the type in use from the end of the 11th century, which has been discussed above.

It appears that in Norway during the 12th

---

13 Öland, the second largest of Sweden's islands, lies off the coast of Småland in the south east of the country.
14 Julin was situated on the island of Wollin in Pomerania, now eastern Germany, on the site of the earlier Viking fortress of Jomsborg. The power of the Jomsborg Vikings was broken in the battle of Hjörungsväg in 986 (see p.17).
15 A literal translation of Alm's word *ledningsskep* would be 'command vessel', but this makes no sense. It seems likely he meant *ledungsskep*. The Ledung was an organisation for military service at sea in the early middle ages. This would suggest the translation given of 'naval vessels'.
16 On the death of Magnus Barefoot in Ireland in 1103 his sons Sigurd Östen and Olaf ruled Denmark jointly. In 1122 Sigurd became sole king and reigned until his death in 1130.

century little attention was paid to the papal ban on the use of crossbows against Christians. Crossbows were in very common use in the civil wars of the time (108:129). King Sverre himself, and his Earl Philip, shot with 'lock-bows' at a battle in the Trondheim Fjord in 1199 (108:175). Their crossbows certainly had composite bows, and were probably of English manufacture or English type, for King Sverre enjoyed very good relations with King John of England, who among other things sent him English bowmen (108:194). During the 12th and 13th centuries English trade with Bergen was generally vigorous, and English culture had a considerable influence in Norway.

In the Norwegian treatise, *Konungs Skuggsja*, (A Mirror of Kings), probably dating from the middle of the 13th century, *låsbågar* [literally 'lock-bows'] are often mentioned (62:140). Mounted soldiers were advised not to use bows so powerful that they could not be spanned from horseback (62:139-140). Young men were urged to *keep busy with their target shooting* with lock-bows or hand-bows (62:135).

There is no specific evidence about Swedish crossbows from the 12th and 13th centuries. However, we may assume that at this time these weapons were quite commonly used in the royal army, by the high nobility, and also in the cities where German influence predominated.

In the Statens Historiska Museum [National Historical Museum], Stockholm, there is a yew bow for a crossbow which dates from the Middle Ages. It was found in Västergötland.[17] The length is 105.5cm, and at the middle the bow is 2cm thick and 7.2cm broad.

During excavations at Aranäs, the citadel of Torkel Knutsson, a crossbow nut of horn was found together with a two-clawed spanning hook, 9.4cm long and 5.9cm broad and another spanning hook 20cm long with a single claw or tongue which had been broken off.[18] In addition, a number of heads for crossbow bolts were excavated. All of these have sockets and the majority have short tips of rhomboidal cross-section. Most of them are about 6.5cm long (88:58-59). Aranäs was destroyed by a great fire, probably shortly after the chieftain was taken prisoner on 6 December 1305 (88:87,88).

---

17 The region of Västergötland lies west of Lake Vättern in central Sweden.
18 Aranäs Castle lay on the shores of Kinnekulle in Västergötland. It was the castle of Torkel Knutsson, Marshal of King Birger Magnusson (ruled 1290-1318, until 1298 as a minor). The castle was probably destroyed in 1307.

3 The 14th century

In the greater part of Europe during the 14th century the crossbow was the most important projectile weapon both for warfare and hunting. During the Hundred Years' War the famed English longbow proved to be an effective competitor, something which cannot be said, on the other hand, of the hand firearms that made their appearance in the middle of the century, which were very limited in effectiveness and had a very slow rate of fire.

In western and southern Europe crossbows of the type used in the 13th century continued in use for a long time, both those with a one-foot stirrup and those with a two-foot stirrup.

The great longbow reigned absolutely supreme in the English field armies of the 14th century. However, in towns and castles that were in English hands extensive use was made of crossbows, as in other European countries. Thus, for example, in 1301, Edward I ordered that 12 crossbows with two-foot stirrups and 3000 bolts for them, as well as 5000 bolts for crossbows with one-foot stirrups, should be sent to the town of Linlithgow (75:60, note 1). In 1307 King Edward directed that for the war with Scotland, 100 crossbows with one-foot stirrups and 40 crossbows with two-foot stirrups (*100 balistas unius pedis, 40 duorum pedum*) should be made (63:175). In 1328 Edward III ordered that among other things 100 *arcubalisti ad Pedem*, should be supplied for the defence of the Channel Islands (75:60, note 1). According to inventories of 1344 and 1366 there were at Dover 126 crossbows, 34 of which had composite bows and two-foot stirrups, and nine of which had composite bows and one-foot stirrups (63:176, note 5). In the English army accounts for the years 1372-1374, 49 crossbows are listed, eight with composite bows and the rest with wooden bows (63:179, note 5).

The Burgundian military accounts of the latter half of the 14th century provide very detailed information about crossbows. In a list of 1362, 189 light and heavy crossbows with composite bows and 382 light crossbows with wooden bows are included (79:658). In the same year crossbows with one-foot stirrups (*à un pied*) and crossbows with windlasses and two foot stirrups (*à tour et deux pieds*) are mentioned (79:657). Crossbows with one-foot stirrups were spanned with the spanning-belt (*baudrier* or *baudré*), presumably using a spanning hook, but there is no indication of the sort of windlass with which the crossbows with two-foot stirrups, referred to above, were spanned. It was perhaps a more primitive form of the English windlass illustrated in figure 20. However, Payne-Gallwey supposed that what was involved here was a hook attached to a block (75:73, note 1), and this does seem possible in this case (figures 25 and 26).

According to a contract of 1384, in Burgundy a complete crossbow in the Genoese fashion with iron bands (*à liens de fer*) cost 20 *gros tournais* of silver. The 'iron bands' probably refer to the reinforcement of the stock on either side of the lock. A crossbow with a two-foot stirrup, commanded the same price (79:658). A crossbow with a one-foot stirrup, complete with string, trigger, foot stirrup, and requisite accessories cost 15 *gros* in silver (79:657). The strings were manufactured of thread or yarn from Antwerp (*fil d'Anvers*) (79:660).

In France crossbows of the same types as in Burgundy were in use. According to an ordinance of 1351 every French crossbowman was to have, amongst other things, a good crossbow, the spanning power of which matched his physical strength, together with a good spanning-belt (63:101, note 1). In the mid-14th century the French were using large forces of Genoese crossbowmen. Each of these crossbowmen is reported to have taken the field with only twelve bolts (19:405), which seems to be a very small number. The Genoese suffered a severe defeat at the hands of the English longbow-men at Crécy in 1346. This defeat was reported to have been occasioned by the fact that the crossbow strings of the Genoese had been drenched by a downpour immediately before the battle. A wet bowstring is, of course, useless. The English protected their bowstrings against the rain by removing them (84:50-51). If the strings of the Genoese were indeed drenched by the rain, they must have been of very primitive construction and the rainfall must have been uncommonly severe. It is said that the Genoese had crossbows with composite bows which had slack strings (75:5-6), but this seems utterly incredible. Payne-Gallwey immersed a crossbow, the string of which had been painstakingly impregnated with wax as described above, in a barrel of water. After 24 hours the string had not absorbed any moisture and was entirely fit for use (75:5). The story about the wet strings is probably a Genoese explanation of the defeat intended for home consumption.

In 1321 a distinguished Venetian, Marino Sanuto, called Torsello, delivered to the pope a fascinating proposal for a crusade. In it he gave a detailed description of the weapons and armaments that would be required. In the chapter on the crossbow he mentions ones with wooden bows and says that crossbows with composite bows were better in dry areas than in countries with humid climates. He also refers to crossbows with two-foot stirrups (56:636, 640; 63:113, note 2).

According to an ordinance of 1369, the crossbowmen who formed part of the forces of the city of Florence were equipped with crossbows, spanning hooks and such like (63:98).

In Spain crossbows were used a great deal during the Middle Ages, but information about their design and construction has not been available for the present study.

In Germany the crossbow was very popular in the 14th century. Throughout this century, two types of hand crossbow were in use, the stirrup crossbow (*Steigreifarmbrust*), subsequently called the foot-loop crossbow, and the back crossbow (*Ruckarmbrust*). In 1307 and 1308, ten back crossbows called *balistas dorsales*, and ten foot-loop crossbows (*balistas stegerepas*) were purchased for the city of Hamburg. According to the accounts of Greifswald, from 1361 to 1363 *balistae dorsales* and *balistae strepales vel stegherpes arborstes* were purchased (18:74-75).[19]

In the 14th century accounts of the Teutonic Knights, which go back to 1364, back crossbows (*rugarmbrost*) and stirrup or foot-loop crossbows (*stegereyfarmbrost*) are included. Around 1390 the back crossbows are in the majority, but later on the foot-loop crossbows became increasingly numerous. In addition, a distinction is made in the accounts between new and old, large and small crossbows, and between *Gemeine-*, *Knüttel-*, *Diener-*, *Gesellen-*, and *Schützenarmbrost* (79:364).[20]

The difference between the back crossbow and the foot-loop crossbow is nowhere made clear and almost every scholar has his own theory about this. The Danish arms historian Otto Blom considered that the foot-loop crossbow was spanned by using both hands, and the back crossbow by using a spanning hook (18:74-75). The German Köhler held that the foot-loop crossbow was the same as that known in Western Europe as the crossbow with stirrup for one foot, that *rück* [*rücken*] (the back) was a distortion of the French *croc* (hook), and that the back crossbow was a crossbow with a stirrup for two feet, intended to be spanned with a spanning hook (63:117). This appears entirely credible. Finally,

---

19 The town of Greifswald stands on the Baltic Coast of Pomerania, now eastern Germany, some 50 miles east of Rostock.
20 These German terms can be translated roughly as follows: *ordinary crossbows; stave crossbows; crossbows for the use of servants, crossbows for the use of companions, fellows or members of the order; and target crossbows.*

the German Rathgen says that a back crossbow is spanned with a goat's-foot lever (figure 15) and that the foot-loop crossbow is spanned with a spanning hook (75:654).

In Germany in the 14th century wooden and composite bows appear to have been equally common. The 14th century ordinances relating to the imperial hunting reserves contain interesting evidence. Thus, for example, one dating from 1382 instructs the appropriate hereditary master of the hunt to deliver to the Emperor, on the occasion of his visit, a crossbow with a bow of yew, a tiller of maple, a nut of ivory, and a string of silk (64:37). In 1396 7000 bow-staves of yew (*ywen bogenholcz*) and 1150 of *knottelholcz* were sent from Ragnit to the Grand Master of the Teutonic Order.[21] Rathgen supposes that these bowstaves were intended for the manufacture of the simpler classes of crossbows, the *gesellen-* and *knüttelarmbrost* (79:655).

According to the accounts of the city of Hamburg, in 1373 *ene zule* and *una statua ad balistam* were procured, and in 1375 the city paid '*vor ene armborstes zule*' (18:76). The Latin word *statua* and the Low German *zule*, in High German *Säule* (19:405), mean the tiller of a crossbow. The Low German word was used in Sweden as well, certainly throughout the fifteenth and sixteenth centuries and was quite probably used earlier as well. In Sweden it assumed the form *sula* (77:68, note 1), *armborstsula* (52:233), *sull* (2:225), or *sool* (12:149).

In the lands of the Teutonic Knights crossbows had reinforcements and inlays of stag or elk horn, which were often decorated. The strings there were made of Flemish yarn (79:665), which presumably means the same as the *fil d'Anvers* of the Burgundian accounts (79:660).

During the 14th century most German crossbows appear to have been spanned with the spanning-belt and the spanning-hook. Such belts are referred to at Frankfurt am Main, where they were manufactured of oxhide and were fitted with hooks and rings (79:645, see figure 14). In 1372 the city of Hamburg purchased *twe spanhaken* (18:76). Spanning belts with hooks are mentioned at Trier in 1379

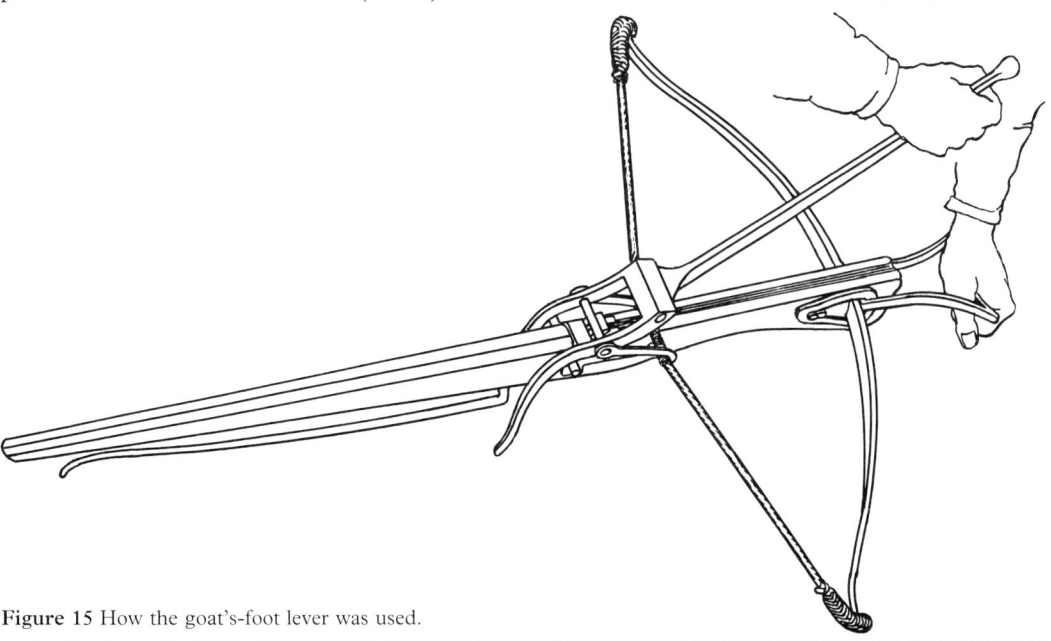

Figure 15 How the goat's-foot lever was used.

---

21 Ragnit is a town on the River Nieman, or Nemen, just upstream of Tilsit. This river has historically marked the boundary between East Prussia and Lithuania. The word *knottelholcz* has so far defied translation.

iron, and to these one or two movable iron hooks were attached. Eventually the whole device came to be made of iron (figure 15), constructed in such a way that when folded the handle lay between the tines of the fork (figure 16). The bender is in this latter case almost universally equipped with a broad hook so that it can be carried on the belt. Figure 15 shows how the goat's-foot lever operated in spanning. The claws of the fork were laid against the projecting end of an iron lug that passed through the tiller, a little behind the lock. When the handle was then drawn backwards, the two hooks pulled the string into the spanned position.

In 1395 Stockholm Castle was besieged by a force of Hanseatic troops. The contingents to be supplied by Prussian cities and the armament for the men at arms were determined at a Hanse congress on 12 July 1395. The cities were to send a joint total of 41 squires and 30 crossbowmen to Stockholm. The city of Thorn was to send not only men at arms but also a crossbow maker and, in addition, like Elbing and Danzig, a crossbow windlass, four benders, three *ader* (crossbow strings), and ten windlass crossbows (106:1-2), or one windlass crossbow per crossbowman.[23] According to the same ordinance, each crossbowman was to bring with him a large, a small, and a medium-sized crossbow (106:1-2).

It is doubtful whether the windlass crossbows referred to above were really what one might call rampart crossbows. More probably they were hand crossbows of the largest size. However, true rampart crossbows, which were also called windlass crossbows, and which were set upon stands or mounted in carts for use had appeared by this time throughout almost the whole of Europe. If they were of the largest class and were mounted on carts they were called *springal, springol, springarde, espignol, ribald*, and the like (63:177-178). They could

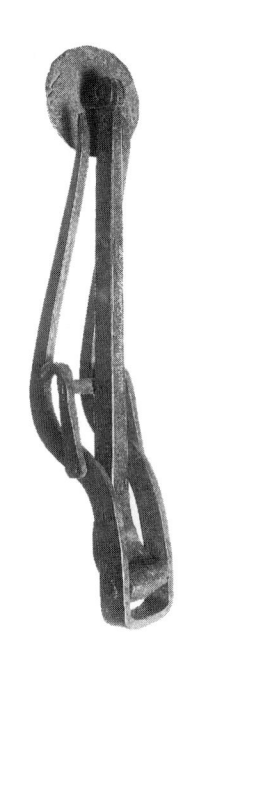

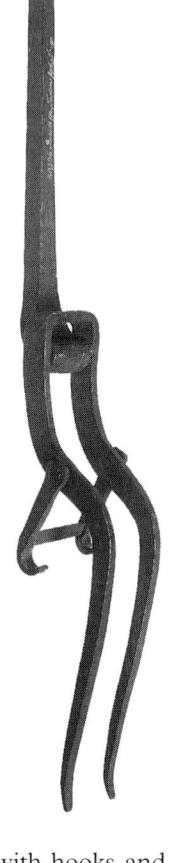

Figure 16 Goat's-foot levers. To the left - folded; to the right - ready for use. Livrustkammaren [Royal Armoury], Stockholm.

(79:650), and spanning-belts with hooks and quivers were purchased in the lands of the Teutonic Knights in 1399 (32:229).

In the 14th century a new spanning device, the 'goat's foot lever' (*getfoten*) also called the 'bender' (*vippa*), came into use.[22] In principle it was a lever with a single arm. The spanner used for the Norwegian whale bows, described above (figure 3), shows the goat's foot in its simplest form. Later on the handle appears to have been made of wood, the forked claws of

---

22 A literal translation would be 'pump-handle' or 'lever'. The term 'bender' has been adopted as standard in this translation as it is understood by most students of the crossbow.
23 Thorn is a town in West Prussia (now Poland) on the right bank of the river Vistula, approximately 100 miles south of the Baltic. Danzig (now Gdansk) lies on the Baltic coast of Poland. Elbing (now Elblag) is a town some 22 miles south east of Gdansk, near to the coast.

be of considerable size. At Freiburg a bow for such a crossbow has been found: it was 13ft long.[24] However, crossbows are said to have been found with bows as much as 6m long (63:179)

In England rampart crossbows are mentioned in 1301 (75:60, note 1), in 1307 *balistas a turno*, and 1328 *arcubalisti ad Troll* (75:60, note 1). At Dover in 1344 and 1366 there were three large windlass crossbows (*iii magne arbaliste ad turno*) (63:176, note 5). In 1362 rampart crossbows (*grosses arbaletes*) are mentioned in the Burgundian military accounts (79:657). Rampart crossbows were also common in Germany. In 1307-8 the city of Hamburg purchased four *Wintarmbruste*, which were also called *magnas balistas* (18:75). In the accounts of Greifswald for 1361-3 purchases of a *Wintarmborst* are mentioned (18:74). In 1378 a large crossbow was purchased for Trier from Luxembourg: it cost more than four times as much as an ordinary hand crossbow (79:650). In the lands of the Teutonic Knights the windlass or rampart crossbow was frequently called the *bankarmbrost*, after the bench (*bank*) upon which it was placed for shooting. In the 14th century windlass crossbows appear mainly in the oldest of the Order's accounts, which begin in 1364. At 30 of the 54 castles, fortresses, and cities mentioned in the accounts no such weapons were found (79:654)

In the Schweizerisches Landesmuseum, Zurich there is a fairly small rampart crossbow, (figure 5), which dates from the 14th century. Its primitive lock is described in the first chapter. The weapon has an iron stirrup and it may well represent a very common type both of rampart and hand-crossbows of the 13th and 14th centuries.

The windlasses or tackles used for rampart-crossbows are quite frequently referred to in accounts of the time, but always without any details of their design. During the latter half of the 14th century they appear to have had cranks rather than spokes (63:182, note 4), and they were presumably much like the English windlass (figure 20). In 1367 at Greifswald there is mention of an *'instrumentum balistarum dictum armborstwinda'* (18:74). In 1372, along with other things, the city of Hamburg purchased *'ene winden'* (18:76). In the lands of the Teutonic Knights there is one mention of a crossbow windlass of wood (79:654).

Every German city had a permanent, official crossbow maker, whose job it was to make a certain number of crossbows for the city every year. The title and the conditions of employment varied from town to town. From the beginning of the 14th century until well into the 16th Hamburg had a permanently employed craftsman of this type who was required to produce four crossbows annually. For any further pieces delivered he was paid extra. In the accounts he is most often called *balistarius*, but in 1370 and 1374 *balistifex*, and many times, for example in 1356 and 1386, *Armborstmakere* (18:78). In Naumburg the city crossbow maker was usually called the *Schützenmeister*; but in 1349 and in the following year he was called *sagittarius*, and in 1371 once *sagittarius* and once *magister ballistarius* (79:647). In Görlitz the city's *ballistarius* is mentioned for the first time in 1364 (79:651).[25]

In the 14th century war bolts frequently had sockets and short, heavy tips of rhomboidal cross-section. Long dart-like tips with tangs were also used. In the High German of the Middle Ages the shaft of the bolt was called the *zeyne* (79:644, note 3), and in Switzerland the *pfilzeine* (118:45), which has become *zaine* in modern German (19:424).[26] The Low German name in the Middle Ages was *thene* (18:76), a name which is used in the Middle Ages both in Denmark (18:76, note 2) and in

---

24 Freiburg lies on the western side of the Elbe estuary some 40 miles north west of Hamburg.
25 There are two towns called Naumburg; one is in western Germany, some 15 miles west of Kassel; the other in eastern Germany on the river Saale, some 30 miles south west of Leipzig. Gorlitz, too, is a town in the south east of Germany on the border with Poland some 50 miles east of Dresden.
26 In modern German the common meaning of *zain* is ingot, or bar. *Zaine* is a dialect word for basket- or wicker-work.

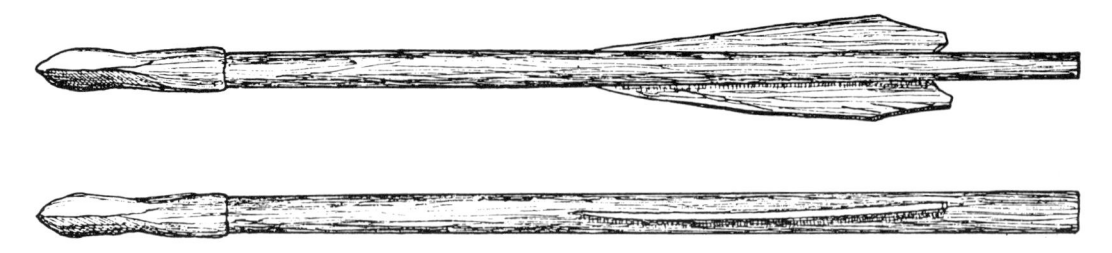

Figure 17 Crossbow bolts with socketed heads and spirally-set flights. International type, 14th to 16th centuries. Length 39.7cm, weight 87g. Livrustkammaren [Royal Armoury], Stockholm.

Sweden, where shafts were mostly called *piltenar*.

In order to ensure that the bolt could be shot both accurately and a long distance, the centre of gravity had to be in the proper position. For bolts which were 35cm, long or shorter, the centre of gravity was usually exactly one-third of the way along the shaft, but for longer ones it was a quarter of the way along from the head. The position of the centre of gravity of each bolt was carefully checked and was regulated by cutting a little wood out of the shaft at a suitable point (19:424). It was usual for the rear of the shaft to be of the same thickness as the string, usually 1.2-1.3cm (75:17).

The flights might be made of real feathers, parchment, delicate slices of wood, or thin sheet copper, the last being used especially for the bolts of rampart crossbows. Usually there were two flights, sometimes three. For the most part they were straight, but they could also be set spirally, which gave the bolt a rotating motion and appreciably improved accuracy (19:427). Bolts of this sort (figure 17), usually have two flights. In 1349 400 such bolts (*drelinge*) were made in Frankfurt am Main (79:646).

During the period when the crossbow was an important weapon, ideas on the usefulness of fletching varied considerably. The heaviest bolts generally are without flights (19:426).

The length and weight of the bolts had to be precisely adjusted to suit the crossbow with which they were to be used. For this reason in western Europe there were two main types; one for crossbows with two-foot stirrups and for windlass crossbows, and another for crossbows with one-foot stirrups (79:659). In the lands of the Teutonic Knights the distinction was primarily between bolts for foot-loop, back, and windlass crossbows. However, for each type, large and small, and old and new, were found (79:655).

It appears that in Germany during the middle and the latter half of the 14th century the types of crossbows and bolts changed. During this period at Frankfurt am Main many crossbows were modernised, and great quantities of crossbow bolts were shortened (79:645).

On campaign the shafts of bolts were carried in kegs or casks (79:646). Because of this it was usual to count large quantities of bolts and shafts in terms of kegs. In northern Germany, Denmark, and presumably in Sweden as well, it was reckoned that a keg contained 800 bolts (18:76, note 2).

During the Middle Ages, at least in northern and western Germany, the term for the process of fitting shafts was *sticken* (79:644, note 3). The craftsman who did this work was called the *pilsticker* (79:648, note 11). He presumably 'stuck' the shaft into the sockets, and the tangs into the shafts. These German names were also adopted in Sweden, where they became *sticka* or *stycka* and *pilstickare* (95-97). The 'shaft-stickers' also manufactured the flighted shafts (79:644, note 3).

Every crossbowman in the contingent sent by the Prussian Hanse towns to Stockholm in 1395 was to have a shock (sixty) of good, socketed (*getulleter*) bolts. Thorn, Danzig, and Elbing were each to send two kegs of bolts. There are no further details regarding the bolt

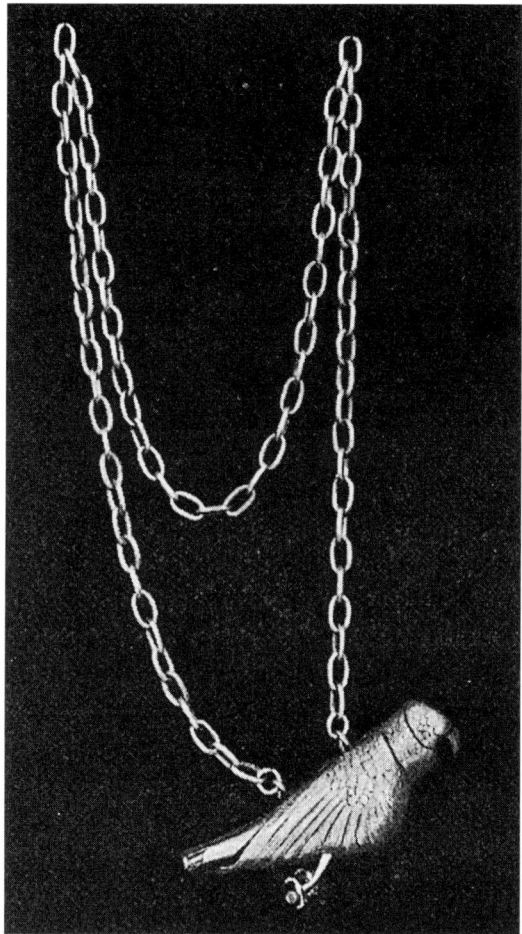

Figure 18 Parrot with chain of gilt silver. Challenge prize at popinjay shoot in Visby, end of 15th or first half of 16th century. Photo, Livrustkammaren [Royal Armoury], Stockholm, from Rudolf Cederström's archive. Gotlands Fornsal [Museum of the Province of Gotland], Visby.

called the best and most popular form of target-shooting was shooting at the popinjay, mentioned earlier, and which, by the way, began in ancient times. Such competitions were common in the Middle Ages in northern France, Holland, Belgium, Germany, England, Switzerland (91:326-327), Denmark, and Sweden. In 1354 the Grand Master in Prussia had 'a tree with a bird' set up in every city. The bird was of wood, as big as a hen; it had outstretched wings and was painted in motley colours, for which reason it was called the popinjay or parrot (91:328). The height of the tree, or later the pole, varied from 7 to 17 m. At the top of the pole was an iron bar on which the popinjay was placed. Blunted bolts were used for shooting at the bird. Whoever was fortunate enough to shoot down the bird was dubbed the 'shooting king' and received a sort of challenge prize, a gilded parrot, to be worn around the neck on an accompanying silver chain. The parrot was worn not only during the solemn procession after the end of the shooting-match, and during the guild meeting which followed, but also at all the major festivals up until the next year's competition. If the same marksman won the parrot three times in a row, it became his property, although the guild had the right to buy it back from him (113:254).

The 'shooting king' also received other prizes, often a silver beaker, plus leather or cloth for a pair of breeches, since breeches were subjected to heavy wear in the spanning of crossbows (113:254). On the other hand he had to invite the participants to a festival dinner, which in many cases must have proved to be very costly, in view of the unparalleled ability of the men of the Middle Ages to eat and particularly to drink. For regular shooting competitions the city underwrote at least the ale, and very frequently also the food, consumed by the competitors (79:647, 649, note 13).

From Scandinavia we have only a small amount of evidence concerning crossbows in the 14th century. Nevertheless these weapons were very common there. There were three

deliveries of Danzig and Elbing, but Thorn was to deliver one keg of socketed (*getullet*) and one keg of 'stuck'(*gesticket*) bolts (105:1-2).

The socketed bolts were presumably of the same type as the ones illustrated in figure 17. A 'stuck bolt' in Sweden and western Germany usually meant a ready-assembled bolt, but in this case it probably means bolts with tanged heads.

The German cities, and to a certain extent the nobility as well, laid great emphasis upon training in marksmanship. All cities had shooting ranges. One of these (*zhelstat*) is mentioned at Görlitz in 1377 (79:652). What might be back crossbows (*'balistas quae dicuntur Ryggearmbyrsth'*) in Copenhagen Castle in 1328 (18:74). It is apparent from the injuries to the skeletons of the fallen soldiers that the Danish crossbowmen at the battle of Wisby, fought in 1361, had effective weapons. In a number of cases bolts have gone straight through skulls. All the bolt-heads found have sockets and are 4-6cm long (112:124, 186-187). An inventory of 2 April 1373 of Orum castle in Denmark includes three *balistas*, two dozen *telorum* (bolts), and a crossbow windlass (18:74).[27]

In Sweden the crossbow became increasingly popular during the 14th century. However, it seems to have been used mainly in the towns and by professional soldiers. It does not yet appear to have gained a footing among the peasantry. The Södermanland Laws adopted in 1327 decreed that every man liable to military service was to have, among other items, a handbow with three dozen arrows (109:190), and in the Hälsingland Law, dating from the first half of the 14th century a bow and three dozen arrows are included among the weapons of the common people (47:92).[28]

In 1365 Raven von Barnekow, the Sheriff of Nyköping, bought two crossbows for a total of 14 marks (*'II balistas pro xIIII marchis'*), and paid in all 40 marks to the crossbow maker for these and their accessories (*'magistro balistarum et instrumentis xl marchis'*) (105:100).[29] The unit price for the hundred or so horses that Raven purchased or paid compensation for during the years 1365-7 was $20^{1}/_{2}$ marks per unit (105:C11), and this gives an idea of the value of crossbows at this time.

Alsnö castle on Adelsö island appears to have been destroyed by an assault towards the end of the 14th century, perhaps in connection with the harrowings of the *vitalie* brotherhood in the Mälardal.[30] The grounds there are thickly strewn with crossbow bolts. Six heads have delicate tips and no tangs, but the rest have sockets and are of the usual type (111:50).

At Stockholm the manufacture of crossbows appears to have been in full swing in the latter half of the 14th century, and there is evidence some were exported, including to the lands of the Teutonic Knights where *holmische armbrüste* were highly esteemed. In 1399 crossbows and quivers were purchased at Stockholm for the Teutonic Knights (79:656). Unfortunately all the accounts and records of the city of Stockholm from the 13th and 14th centuries were destroyed in a great fire of 1419. They may well have contained valuable information on crossbows amongst other things. It is possible that *holmische armbrüste* may have meant crossbows of an improved type characteristic of those made in Stockholm.

Crossbows made entirely of wood, of the type described in Chapter 1, were undoubtedly used throughout the whole Middle Ages and for many years thereafter both in Scandinavia and in other parts of Europe, particularly by the less affluent.

---

27 Orum Castle lies near the coast in the district of Hassing in the north west of Zealand.
28 The Swedish region of Södermanland lies on the east coast immediately south of Stockholm. The region of Hälsingland also lies on the east coast of Sweden, but begins some 150 miles north of Stockholm.
29 Nyköping is the chief town of the Swedish region of Södermanland which lies due south of Stockholm.
30 Alsnö Castle on the island of Adelsö in Lake Mälar is mentioned in records between 1279 and 1304. The Mälardal is the strip of low lying land around Lake Mälar. The Vitalie Brotherhood was a group of pirates or privateers who operated in the Baltic and North Seas from 1399 until the mid 15th century. Its ships were based at ports like Rostock and Wismar and plundered Swedish, Danish and Norwegian towns such as Nyköping, Söderköping, Malmö and Bergen. In 1394 the pirates conquered the island of Gotland. In the war beginning in 1416 between Erik of Pomerania and the Dukes of Holstein the Vitalie Brotherhood was allowed to operate from Holstein. Later when Erik had been dethroned and was living on Gotland (1438-49) he let the pirates use the island. Thereafter they are not heard of again.

4 The 15th century

During the 15th century firearms began to drive the crossbow out as a military weapon. As early as the Hussite wars of 1419-34 guns had proved suitable for use under field conditions. In about 1431 it was laid down that one-third of the imperial infantry should be marksmen, and that of these one half were to have crossbows, the other half hand guns (90:407). In the second half of the 15th century hand firearms were preponderant in the armies of central Europe. However, the crossbow continued to be used to some extent for military purposes until well into the 16th century.

The design and construction of the crossbow were improved in the 15th century, presumably as a result of the great competition from firearms. From the middle of the century it became common to drill a hole through the lockplates, the tiller, and the centre of the nut. A cord was drawn several times through the hole and across the under side of the tiller, and in this way the nut was held firmly in place both when the crossbow was being carried and when it was not in use (19:409). On western European crossbows in particular the cord was usually replaced by a rod, or sometimes by a pair of short screws, one of which entered from each side.

Also, in about the middle of the 15th century, a spring was introduced which pressed the forward end of the trigger against the nut (figure 19) (19:409).

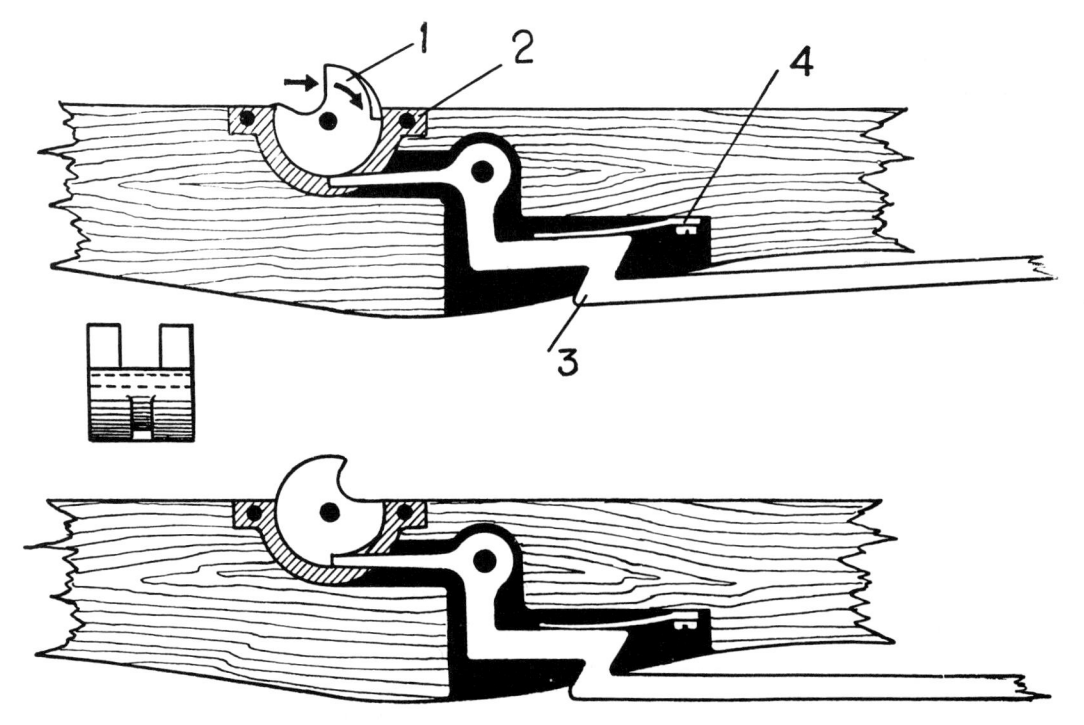

Figure 19 Crossbow lock with trigger spring. Western European type. Above: released. Below: spanned. *1* - nut; *2* - seat for nut; *3* - trigger; *4* - trigger spring. After Payne-Gallwey.

33

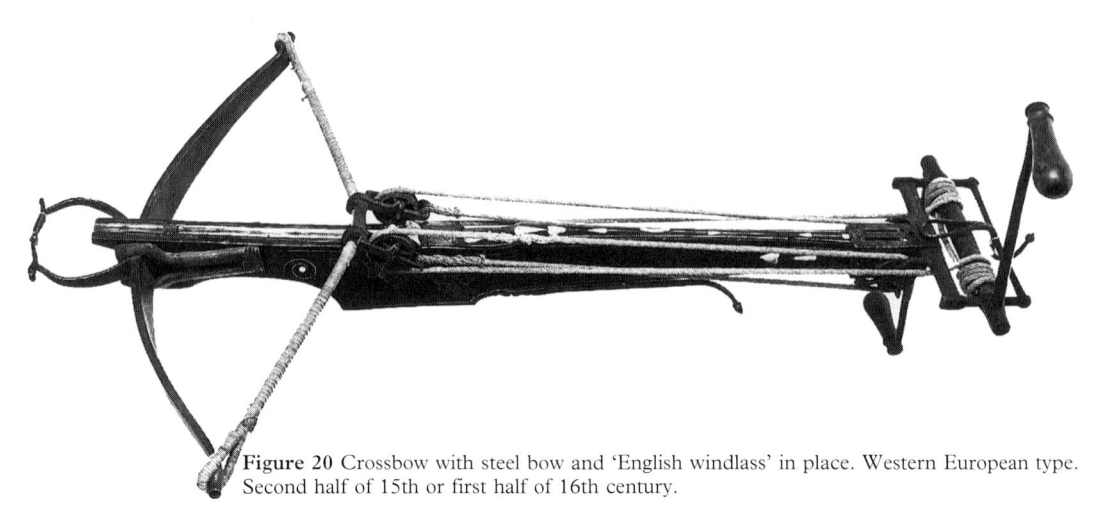

Figure 20 Crossbow with steel bow and 'English windlass' in place. Western European type. Second half of 15th or first half of 16th century.

By the beginning of the 15th century most crossbows had composite bows, but soon these were subject to strong competition from steel bows. Steel bows had been tried out earlier, but it was only after about 1425 that they began to appear in large numbers.

In Burgundy, steel bows were procured for the army for the first time in 1437. Between then and 1448 more than a thousand crossbows with bows of this sort were acquired. In 1446 *'100 crossbows of Genoese fashion with steel bows'* were put on board the Burgundian galleys that were leaving for Rhodes (79:658). In the Burgundian military accounts crossbows with yew bows were still included for 1433, 1437 and 1442. Between the years 1437 and 1448 37 crossbows with yew bows were purchased (79:658). In 1435 there were three large, four small, and three medium-sized crossbows with steel bows in the English armoury at Rouen (63:186).

When the steel bow was generally adopted it came to be attached in a new way to the tiller, by means of bridles. A bridle (figure 20) was placed on either side of the fore-end of the tiller. In the tiller there was a long, rectangular hole usually about 1.5cm broad and 7 to 8cm long, which started a little behind the tiller's fore-end and ended at the inner edge of the stirrup-like bridle's rearward loop. At the front and back of the hole an iron liner was fitted, its ends being bent forward and back respectively. Between these liners wooden wedges were driven to hold the bridles fast to the tiller and with them also the stirrup and the bow.

Western European crossbows are characterised not only by these bridles, but also by having angular tillers and stirrups, iron sideplates around the lock, and bolt channels which are usually reinforced with staghorn.

Probably as a result of the increasing dominance of the steel bow, it became common, first in Western Europe, to extend the upper surface of the tiller and the bolt-guide forward of the bow, and often the under side of the tiller too. Thus the bow came to lie in quite a deep recess cut into the tiller's forward end.

A crossbow of Western European, type, perhaps from Burgundy, and dating from the 15th century, which was found in Western Switzerland and is now in the Schweizerisches Landesmuseum in Zürich, is very much like the one illustrated in figure 20. It has a stirrup with an angular outline and an inward-curving foot-rest, a pear-wood tiller, and a bolt groove. Behind the lock the tiller is of octagonal cross-section. Beneath the lock the tiller has a reinforcement which provides a good grip for the left hand and which is also convenient if one wishes to rest the weapon on a wall or something similar. The sideplates are of iron, handsomely formed. The lock has a lever-trigger

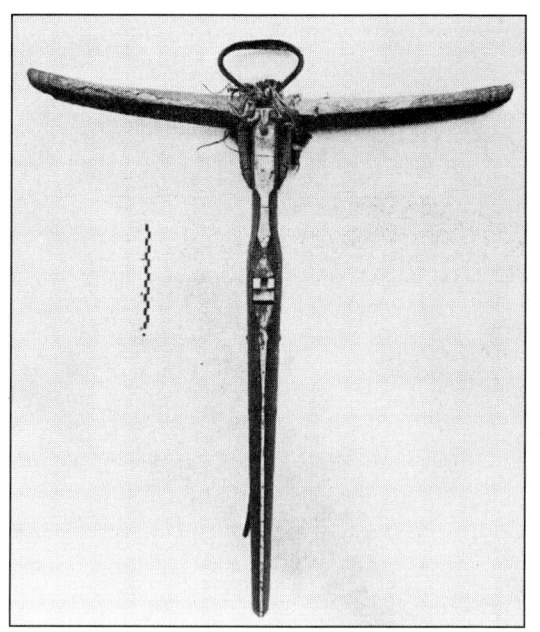

Figure 21 Crossbow with composite bow. South German or Austrian. Second half of 15th century, Bernisches Historisches Museum [Historical Museum of Bern], Bern.

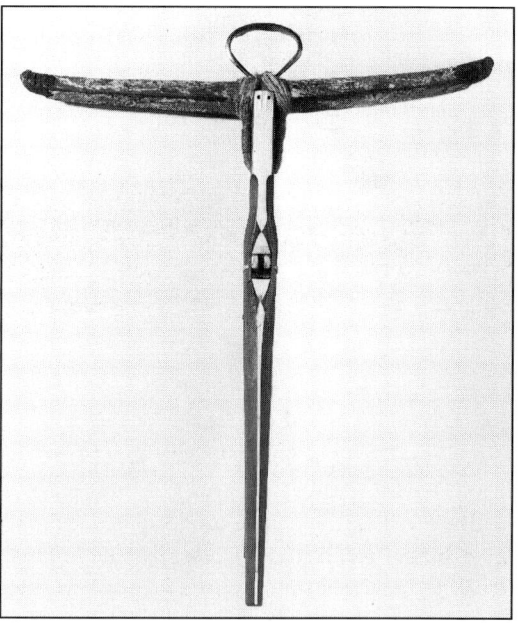

Figure 21a Crossbow with composite bow. South German or Austrian. Second half of 15th century, Bernisches Historisches Museum [Historical Museum of Bern], Bern, Inv. no.237c.

and, as a safety-catch, a small arm fastened to the tiller's under side and pivoting around a pin, which can be set in a vertical position so that it prevents the trigger from being drawn up against the tiller. The safety-catch is released by bringing the arm up against the lower side of the tiller, thus allowing the trigger to move freely. The overall length of the weapon is 104cm, the tiller is 93cm long, and the bow 77cm long. The weight is 6.77kg. The weapon is spanned by means of an 'English windlass', figure 20 (34:390-391).

Crossbows equipped with steel bows soon came into quite general use in Germany, but no more detailed information has been published. In that country and in Austria composite bows continued to be popular for certain types of crossbows for many years, particularly for hunting crossbows. Such bows were regarded as being better than steel bows in very cold weather. The worry was that under such circumstances a steel bow might easily snap (19:405).

A Frenchman who travelled in Bavaria in 1455 said that the Bavarians were good crossbow-shots and that they used crossbows with bows made of horn and sinew. These bows, he said, did not break in cold weather; instead, the colder it was the stronger they became (75:64). However, from the 16th to the 18th centuries Swedish and Finnish hunters do not seem to have experienced any problems with their steel bows when hunting in winter.

There is a limit to the power of a steel bow made in manageable proportions, and it was not difficult to make a composite bow that was as strong as a very powerful steel bow. Nevertheless, steel bows were easier to make, and this is the reason for their popularity. Composite bows of the second half of the 15th and the first half of the 16th century seem in general to have been thicker and stronger than those made at other times. This was presumably due to competition from steel bows.

The crossbows used in the armies of the Teutonic Knights in the 15th century generally seen to have had composite bows. Of the 97 crossbows that were included in the Grand

Figure 22 Crossbow with composite bow. Nut not original. South German or Austrian. Quiver covered with badger skin, and crossbow bolts. Second half of 15th century. Schweizerisches Landesmuseum [National Museum of Switzerland], Zurich.

Master's armoury in 1448, only one had a steel bow. Windlass, stirrup, and back crossbows are constantly referred to in the Order's accounts of this period. After 1410 stirrup crossbows were in the majority. Windlass crossbows appear to have gone out of fashion, and because of this those that were not already portable were altered into hand weapons. In 1434 24 windlass crossbows were sent to Thorn to be made into stirrup crossbows (79:654-655).

Central European crossbows of the later Middle Ages are characterised by tillers and stirrups of rounded form; by bows and stirrups that are secured to the tillers by lashing; by sideplates made of stag or elk-horn; and by their tillers very often being inlaid with the same sorts of horn. Sometimes either a rather broad splint of horn is fitted into the upper side of the tiller ahead of the lock, or this part of the tiller is sheathed with a strip or horn. The splint or strip is flat for the greater part of its length, but towards the front there is a short raised portion in which is a very short bolt-groove.

A south German or Austrian crossbow with a composite bow, dating from the second half of the 15th century, and now in the Bernisches Historisches Museum, [Bern], has a tiller of cherry-wood which is 76cm long. A strip of horn is set into the upper side of the tiller; at the fore-end it covers the full width of the tiller, but it narrows towards the rear. Towards the fore-end it is raised and has a short bolt-groove. The lock has a nut made of staghorn, which is held in place by a cord binding. The long, lever-trigger lacks a spring. Behind the pin on which the trigger pivots, a piece of horn is let into the under side of the tiller above the trigger. Wegeli believed this was a safety-catch that could be slid down into position. The stirrup is of rounded but compressed form.[31] The bow is 72.5cm long, and at its centre it has a circumference of 17cm. The belly is clad with brick-coloured birchbark, and finally with an overlay of strong paper which has been glued firmly into place. The paper is decorated with stamped ornament and figures, among which are the arms of Austria and of the Counts Montfort whose seat was in the Vorarlberg.[32] The crossbow weighs 3.12kg (115:52, No.1777). In the Bern Museum and in other Swiss museums are similar crossbows, (figures 21 and 22) which would seem to be military booty from the so-called Swabian War of 1499. These crossbows were spanned with a spanning hook and belt.

Large rampart crossbows were still in use in the middle of the 15th century, but they were soon to be superceded. In 1435 there was a considerable number of these large weapons in the English armoury at Rouen (63:186). To judge from the Burgundian inventories which list the armaments of cities and fortresses, the rampart crossbows used at this time were mostly old weapons. In 1440 and 1445 the towers of the city walls of Dijon contained various rampart crossbows (*grosses arbalestes*) with bows of yew. One of these is called '*une vielle arbaleste gemelle*', a weapon designed and built to shoot two or more bolts simultaneously. In one of the towers there were two crossbows with wooden bows and one with a steel bow. Along with the crossbows, the inventories include windlasses of various designs. A little later, in the middle of the 15th century, all Dijon's rampart crossbows were replaced by firearms (79:659).

In the armoury of the Swiss city of Solothurn there is quite a small rampart crossbow which dates from the 15th century (figure 23). The overall length of the weapon is 127cm. The tiller, which is 113cm long, is made of cherry-wood, and has a bolt-groove 43cm long. A little ahead of the lock there is a notch in the under side of the tiller which looks like the remains of a point of attachment for the forward end of the original wooden lever-trigger or a lock of the sort illustrated in figures 1 and 5. The lock has a nut of boxwood and an iron trigger-lever which may be a replacement. The nut is 4cm in diameter and 2.6cm thick. It has no 'fingers', in other words it lacks a groove for the rear of the bolt. It is impossible to say whether this nut is a very old type or is a comparatively late addition to the lock. The recess for the nut is lined, front and back, with ivory.

The bow is of yew, 150cm long; it has an almost elliptical cross-section and is 7.2cm broad and 7.5cm thick at the centre. At the ends the width is 2.4cm and the thickness 3.6cm. As a protection against moisture the bow has been covered with painted parchment,

---

31 The Swedish text is unclear here. A literal translation would be: '*the stirrup is of rounded form but gives the impression of having been pressed together from in front*'. It is unclear whether the compression is intended or accidental.
32 Vorarlberg is a province of Austria lying east of Liechtenstein.

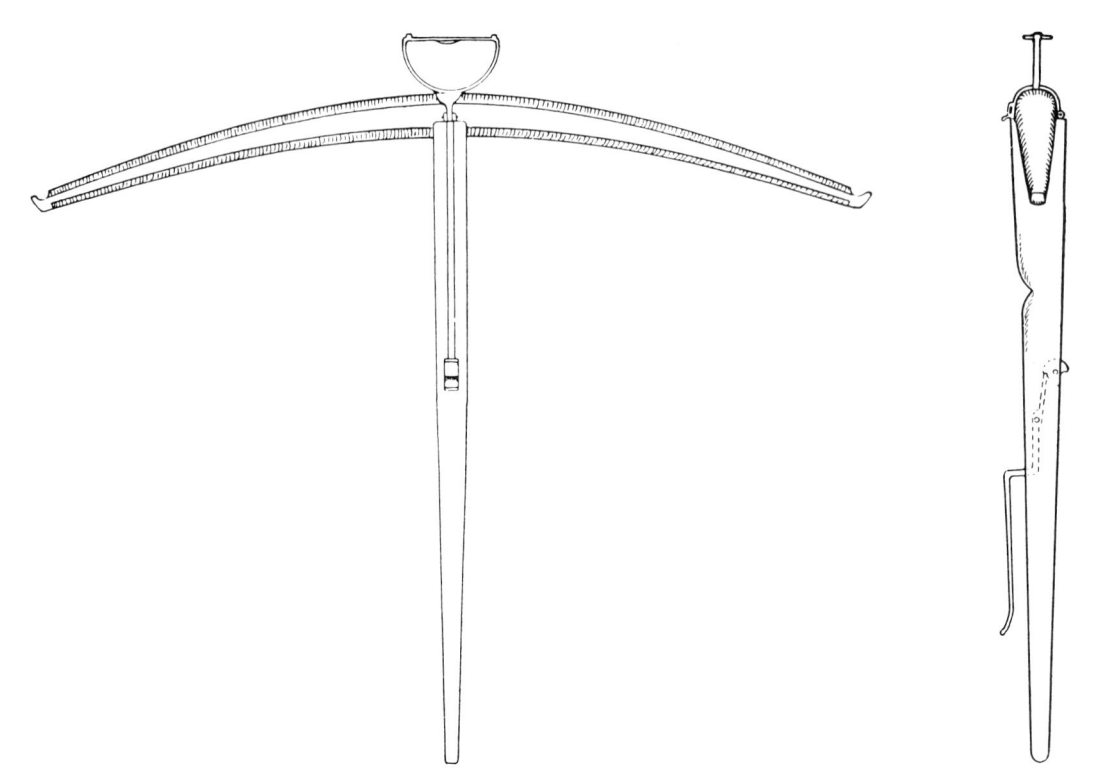

Figure 23 Rampart crossbow with bow of yew. 15th century. Altes Zeughaus [The Old Arsenal], Solothurn.

only very small pieces of which now survive. The bow is fastened to the tiller by means of a peculiar iron bridle, almost semicircular in shape, and thickening towards the centre, which is equipped with a stirrup for the foot. It is attached to the fore-end of the tiller at the top by a hinge and beneath by an iron wedge-pin. To remove the bow one need only drive out the pin and lift up the bridle like a hasp. There is a maker's mark stamped both on the stirrup and on the trigger. The bow should have been precisely fitted to the tiller so that the ends of the bow were a little higher than the upper surface of the tiller.

As I have remarked earlier, the string ought to touch that area only lightly, and should certainly not rub hard against the tiller as it would with the bow in the position illustrated. The crossbow was probably spanned with an English windlass (figure 20). I am grateful to the curator of the Altes Zeughaus Museum, Solothurn for the above information.

Crossbows were made in various strengths and shapes whether their bows were of wood, horn, or steel. It is by no means certain that a steel bow per se would have had greater spanning force than one of horn or wood.

Spanning belts with hooks were used throughout the fifteenth century. Thus, for example, in 1437 'hooks and rings' were purchased in Bern, and spanning hooks mounted upon spanning belts were ordered (118:45). In 1445 12 crossbows with steel bows were bought in Burgundy, and these were spanned by means of hooks (79:658).

In addition, the goat's-foot lever continued to be used to span crossbows of what one might call medium power. Where they were used these spanners seem not to have been made entirely of iron, at least not at the beginning of the century. In 1409 six 'benders' were ordered for the Teutonic Knights, and a locksmith made spanning hooks for them (32:229). At the Knights' fortress of Schlochou in 1420

Figure 24 Cross sections of a cranequin.

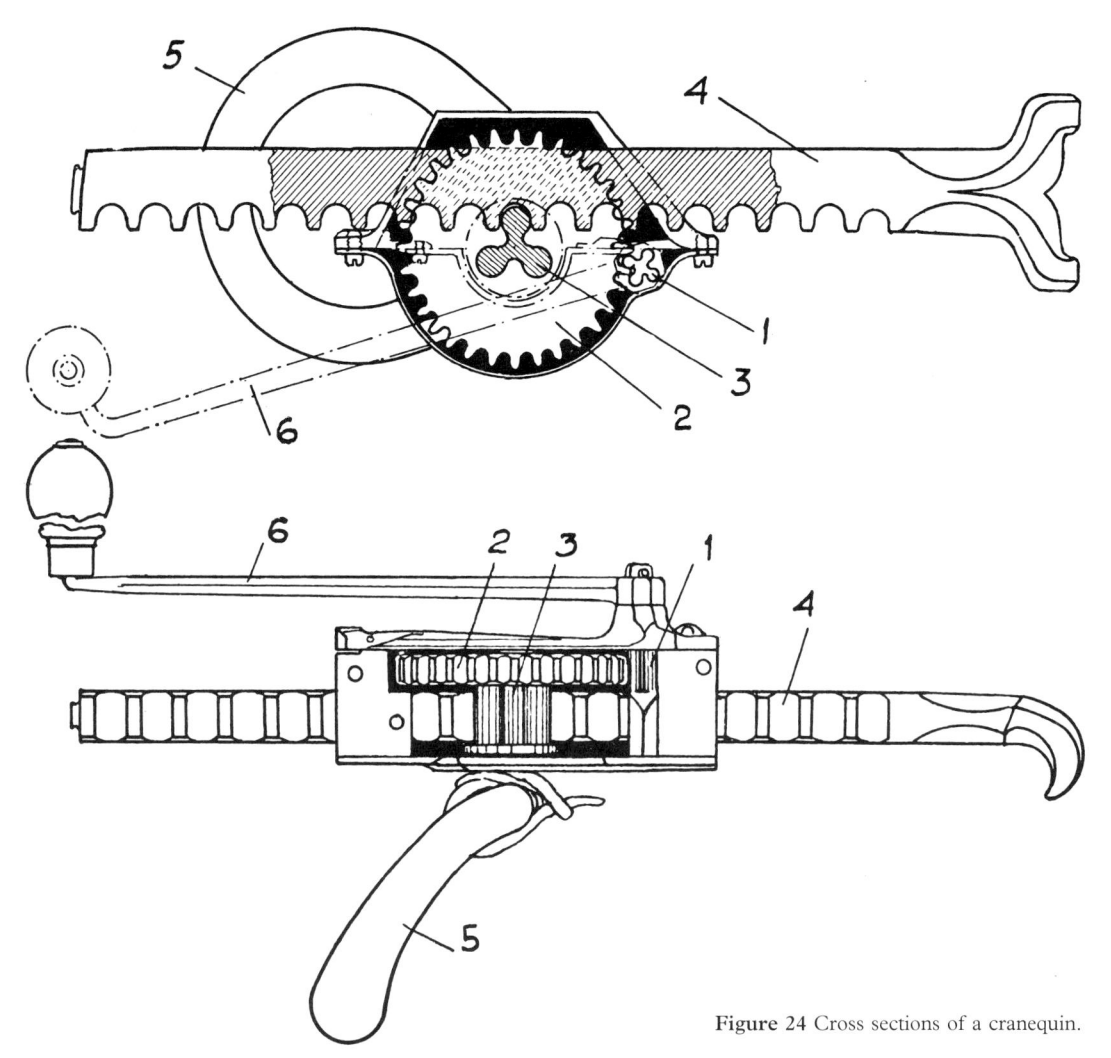

there were twenty hooks (*cropen*) for 'bender' crossbows (79:656, note 23). At Görlitz four benders were purchased in 1422 (79:651) and at Naumburg benders were used in 1448 (79:649).

It seems that 15th century crossbows equipped with very powerful horn or steel bows, used spanning devices of two types - the English windlass, and the rack or cranequin.

The English windlass, (figure 20) is actually a windlass equipped with a winch. It has an open-ended box which is set over the rear end of the crossbow tiller during spanning. In Burgundy the English windlasses, known there by the term *guindaulx*, became very common after 1430. There were two types, simple and double (79:658), the former probably having pulley blocks consisting of two wheels, the latter pulley blocks of three wheels. In 1435 the English arsenal at Rouen contained windlasses of this sort having four, six and eight pairs of pulley wheels in the blocks (63:186). Some of these would undoubtedly have been for rampart crossbows, which, as I have remarked earlier, had for a long time been spanned by such devices before they were adapted, either in the 15th century or a little earlier, to span hand crossbows.

In France crossbows that were spanned by the English windlass were very common around

39

the middle of the 15th century (63:182, note 4).

It appears that the English windlass was used mainly in western Europe, although it is possible that it was also used in western Germany, where there is evidence for a spanning device called *krieg*, which may have been identical to the English windlass. However, like the French *cric*, *krieg* can mean any sort of lifting apparatus, and presumably therefore also any spanning device (63:182, note 5). In 1440 there is mention at Naumburg of a crossbow which was spanned with a *kriegh* (79:649).[33] In inventories of some of the fortresses in the Palatinate three *kriege* are mentioned in 1444 and again in 1449 (63:182, note 5).

The most powerful spanning device was the 'German winch', working on the rack and pinion principle. In Sweden it was known as the *stålbågekran* ['steel bow crank'] (107:39). Incidentally the French name *cranequin* for this rack comes from a Dutch or German word *Kraeneke* (75:131, note 2).[34]

The cranequin is found in two types. In one type the case for the cog-wheel is horizontal, (figure 24), in the other it is vertical. Depending on the type, the wheel case has, either beneath it or at one side, a strong staple to which is attached a loop or sling (5 in figure 24), made of hemp like the crossbow string. Generally at the rear end of the cranequin is a belt-hook. About 12cm behind the lock of the crossbow a steel pin runs horizontally through the tiller. The ends of the pin project quite considerably on either side of the tiller. To span the bow the loop of the cranequin is slipped over the tiller from the rear till it rests against the projecting ends of the pin. The cranequin is then wound forward as far as needed [to engage the string]. Then it is wound back, its claws drawing the string into the spanned position. A cranequin has a very high gear ratio. Joined to the handle (6 in figure 24), is a small

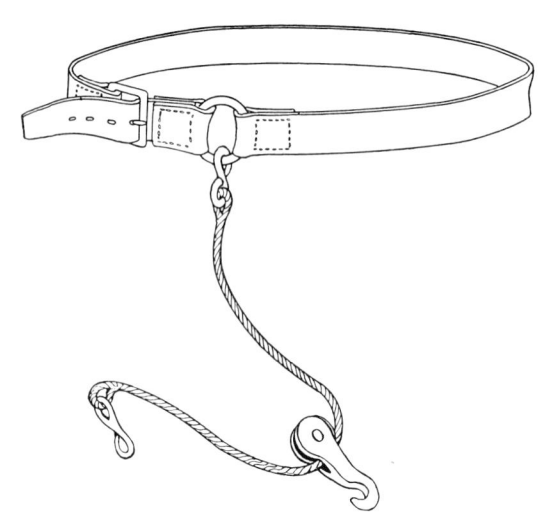

Figure 25 Samson belt, consisting of *krihake*, spanning strap, and spanning belt.

cogwheel, (1) which operates on the large cogwheel (2). Another cogwheel with three cogs, (3) which is, in fact, made in a single piece with the larger one. (2), operates on the rack arm (4).

To begin with, crossbows that were spanned with the rack had a stirrup in which the foot was placed while spanning (115:45). However, later, when these crossbows were made shorter, the stirrup was replaced by an iron ring, by which the crossbow could be hung on a hook or such like.

Even very powerful crossbows could be spanned using a 'Samson belt' which consisted of a belt, a spanning strap, and *krihake*, (figure 25). The *krihake* is a pulley equipped with a hook. Passing through this pulley is a strong braided strap, one end of which is fastened to the crossbowman's heavy belt, and to the other end of which is attached an iron ring. To span his bow the crossbowman put one foot in the stirrup, bent forward, placed the iron ring of the strap over an iron stud or hook set into the under side of the tiller well towards the rear, hooked the claws of the *krihake* over the string, and straightened up, thus drawing the string to

---
33 See note 25.
34 In England in the 16th and 17th centuries this device was known as a rack. However, both because the French term cranequin has been in general use amongst collectors and students for many years, and because the term rack can also be used to describe the racked arm of the cranequin, the term cranequin is used in this translation.

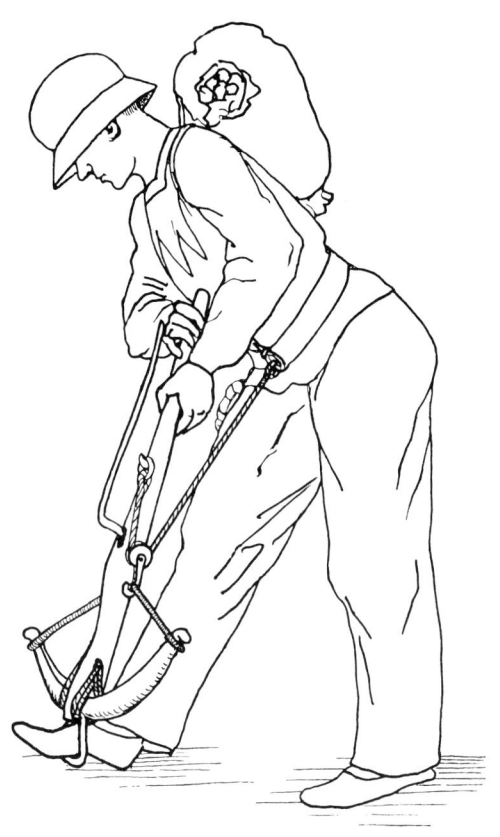

Figure 26 Spanning a crossbow with a Samson belt.

the spanned position using the whole power of his upper body. As Gustav I Vasa put it, crossbows were spanned by '*krihake* and back'.[35]

As I have previously mentioned *krieg* was used to describe any sort of hoisting or spanning apparatus, and therefore it is probable that originally *krieghaken* meant the block in a tackle that was equipped with a lifting hook.

Probably from an early date the *krihake* were made of two paired blocks with hooks (figure 58) which gave better directional control while spanning. This meant that the strap too had to be made double, although with its ends united. Such paired hooks were used with the English windlass and probably a great deal earlier with the windlasses and winches that were used to span rampart crossbows.

Some scholars maintain that the *krihake* is the oldest spanning device designed for the crossbow, older than the spanning hook, and they suggest, although without giving any evidence, that it rarely appears in illustrations later than the first quarter of the 14th century (75:75).[36] It is more likely, however, that the Samson belt was invented, or at least came into fairly general use, around 1400 or perhaps a little earlier.

It is hard to say either where this spanning device was invented, or whether the Samson belt was either a precursor or a simplification of, the English windlass. The name *krihake* might suggest a German origin, but all this means is that the design came to Sweden from Germany. It is a very simple spanning device, but this should not lead one to assume that the *krihake* is older than and a prototype of the English windlass.

In an inventory from the Palatinate dated 1444, there is mention of seven belts with *welle kropfen*, and in another there is an entry regarding three crossbows, three belts, and three *welle kropfen* (63:182). In the accounts of the city of Bern *wellkrapfen* are listed in 1446, and in 1445 there is a payment to a saddlemaker for the assembly of *wellkrapfen* and winches (*'umb wellkrapfen und winden ze vassen dem sattler'*) (118:45). If the *kropfen* and *krapfen* in *wellekropfen* and *wellkrapfen*, were a variant of the usual *cropen* and *krappen*, then *wellekropfen* and *wellkrapfen* could mean a crook or hook equipped with an axle, or in other words, almost certainly, a *krihake*. Simple *krihake* appeared in Italy at least by 1475. (75: figure 31).

---

35 Gustav I Vasa (c.1495-1560) was elected King of Sweden in 1523 and established its independence from Denmark. He introduced the Reformation into Sweden, established strong, centralised government and in 1544 succeeded in changing Sweden from an elective to an hereditary monarchy, thus securing the position of the Vasa dynasty. For Alm's discussion of European references to "back" crossbows, see pp.27-8. For an Arab reference see H.L. Blackmore, *Hunting Weapons*, London, 1971, p.178.
36 Alm took this phrase from Sir Ralph Payne Gallwey, *The Crossbow*, London 1903. In so doing he compounded Gallwey's error. There is ample evidence that the belt and hook spanning mechanism continued in use throughout the 15th century (see inter alia H.L. Blackmore, *Hunting Weapons*, London 1971, p.188).

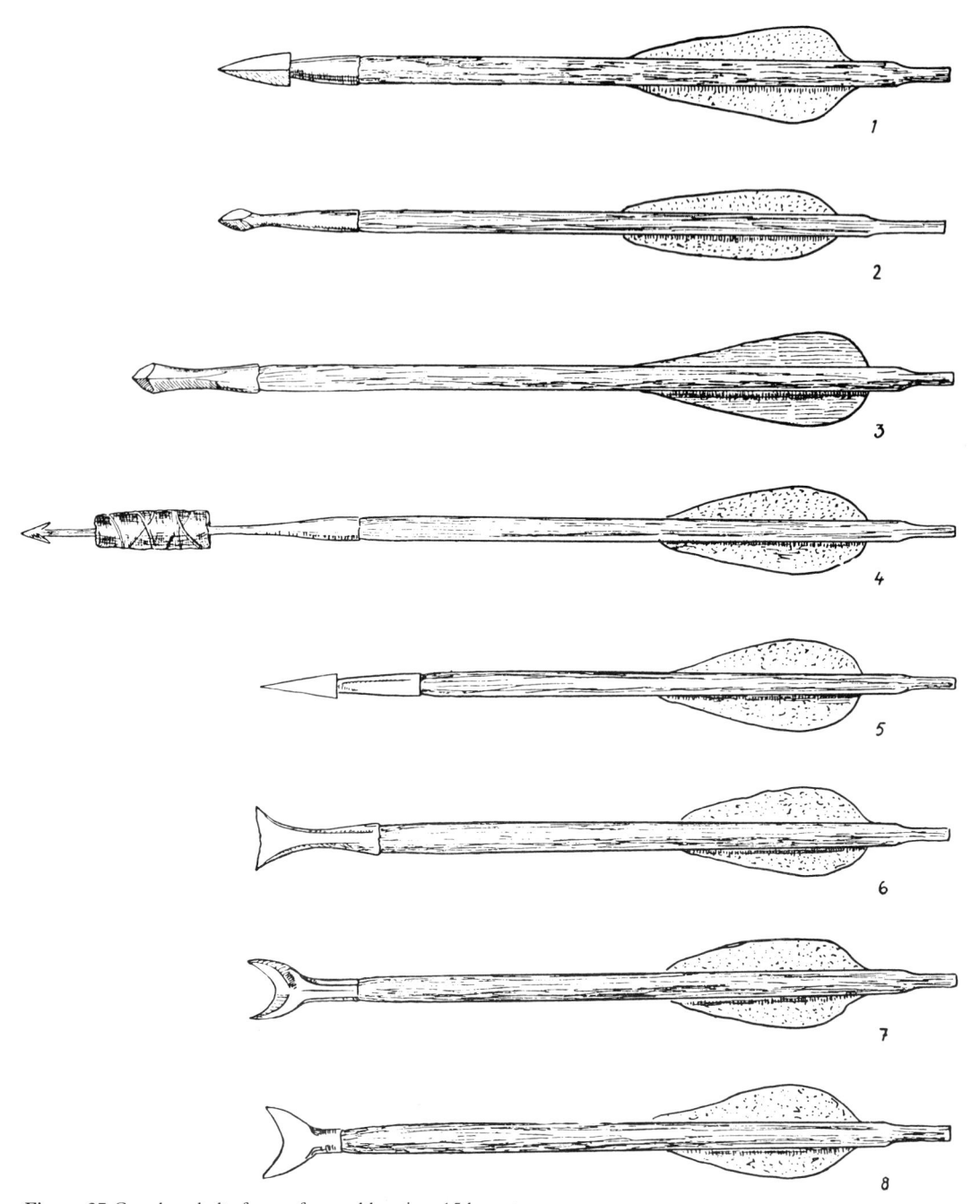

Figure 27 Crossbow bolts for warfare and hunting, 15th century.

In Burgundy in the 15th century very heavy bolts were in use which cost four times as much as ordinary crossbow bolts. In addition, a medium size appeared in 1445 and is found thereafter. The shafts were of oak and the flights were made of slices of wood or thin sheet copper (79:659). In Switzerland, too, bolt shafts were made of oak in the 1440s (118:45). Bolts with spirally-set flights are mentioned in Burgundy in 1417, 1436, 1437 and 1448.

They usually occur in considerable numbers (79:660).

In 1411 the city of Frankfurt am Main had '*4,000 new bolts trimmed to fit crossbow-nuts*' (79:646). Evidently there was a standard way of trimming to fit between the fingers of the crossbow-nut.

The Bernisches Historisches Museum, Bern, has a large collection of 15th century crossbow bolts. Many bolts have heads of the usual socketed type, 6.6-8.5cm long with short points of rhomboidal section (figure 17). They are between 37.8 and 40.2cm in overall length; and they weigh between 58.8 and 79g although most vary only between 70 and 79g. The shafts are of ash, and the two flights, which are slices of wood, are set spirally. On another type the tip has an almost rectangular point (3 in figure 27), while others have an unusually small point (2 in figure 27), and three examples have triangular points of triangular section (1 in figure 27). In these cases the base of the tip is cut straight across to form barbs. Several others have larger tips of the same shape. These last-mentioned types usually have flights made of slices of wood, but occasionally of leather, both types often being set spirally. The overall length varies from 37.7 to 44.7cm and the weight from 40.8 to 49.8g. The bolts with large triangular points of the type illustrated at 1 in figure 27 weigh between 62.8 to 65.4g. However, three similar bolts are equipped with very large tips and weigh respectively 74.1, 74.3 and 72g (116: 37 f).

Incendiary or fire arrows were used throughout Europe. The composition of the incendiary material varied. Usually it consisted of a gunpowder mixture soaked in pitch and oil.[37] This charge was then moulded to a suitable form, and finally dipped in a bath of melted resin and sulphur (116:49-50). In Burgundy in 1431 a batch of incendiary material was made, the constituents being two ounces of camphor, one pound of brandy, two pounds of gunpowder, two pounds of saltpetre, and four pounds of sulphur (79:660).

In the Bernisches Historisches Museum, Bern there are some fire arrows dating from the 15th century (4 in figure 27). The slender, socketed head is 15.2-17.5cm long with a small barbed tip. The shaft is made of ash and the two wooden flights are set spirally. The cylindrical incendiary charge is mounted on the slender head between the socket and the barbs. The only complete fire arrow is 46.2cm long and weighs 105g (116:34-35).

In the territories of the Teutonic Knights there were often large stocks of bolts on hand. Thus, for example, in 1420 at Schlochau there were 63 chests of bolts in the bolt room (79:656, note 23). It was quite usual to reckon bolts by the chest. At Frankfurt am Main, at least, each chest contained between 200-250 bolts (79:646). In addition, according to the inventory quoted above, at Schlochau there were nine sheaves of '*arrows for back-crossbows and also for arquebuses*', 250 sheaves of freshly-shafted bolts, six sheaves of unshafted 'gadflies' (*bromsar, bremsen*) 50 sheaves of feathered bolt-shafts, one and a half sheaves of fire arrows, and eight sheaves of arrows for Hammerstein (79:656, note 23).[38] At Mewe in 1442 there were no less than 1000 sheaves of bolts (79:656) a sheaf consisting of 60 bolts as I mentioned previously.[39]

At this time bolts were still sometimes used

---

[37] Military manuals of the 15th, 16th and 17th centuries give a large number of different recipes for making the incendiary compound for fire arrows. As Alm said, most consist of gunpowder and various additives to make the compound burn longer and more fiercely. Curiously, however, when in 1984 the Royal Armaments Research and Development Establishment at Fort Halstead analysed the incendiary compound on the fire arrow in the collection of the Duke of Northumberland at Alnwick Castle, it was found to be a standard gunpowder mix and nothing more. Until more of the surviving fire arrows are analysed, we shall not know how close to reality were the various published recipes.

[38] Hammerstein lies on the east bank of the Rhine, just over 20 kilometres down stream from Koblenz. *Bromsar* could also be translated as 'horseflies' or 'breezeflies'.

[39] Mewe is a town in West Prussia (now Poland) on the river Vistula.

as projectiles for firearms.[40] The evidence given above that the same sort of bolt was used for back-crossbows and handguns is extremely interesting.

The bolts that were called 'gadflies' at Schlochau, were probably some variant of the Asiatic 'hooters' or whistling arrows which were also used in Russia. In these, either the actual head is drawn out into a sort of whistle-tube, or else, attached to the shaft is a hollow button of bone or metal with a sharp-edged intake hole, which functions as a whistle-tube during the arrow's flight. The piercing whistling noises produced, particularly if great numbers of such arrows were discharged at the same time, must have been very terrifying for the enemy's horses (13:87, 95, 96).

Depending on the purpose for which they were intended, hunting bolts had heads of very different types. It is no exaggeration to say that there was a special model of crossbow bolt for each type of game. Large bolts with quite long, slender heads of triangular or square section (5 in figure 27) were used for hunting stags and bears, smaller ones for grouse and black-cock. Bolts with sickle-shaped heads (7 in figure 27) were used in chamois hunting. These bolts were very effective, but not very accurate. Large bolts wlth broad, crescent heads (8 in figure 27) were used against big game, whereas light bolts with small heads of the same shape were used for duck-shooting (19:428-429). Bolts with chisel-shaped heads (6 in figure 27) were intended for hunting cranes, wild geese, and other larger birds (64:88, note). Blunt-headed bolts (figure 33) were used for hunting small animals for their pelts (19:428-429).

Special tools were used in the mass production of bolts, such as that which took place in the workshops of the Teutonic Knights. In 1417 a smith delivered to Marienburg castle 'two irons to drive bolt-shafts through' ('*2 ysen czu machen czu pfylescheffte durch czu slohen*') (79:656). Evidently what was involved here was the sort of draw-plate still used today for the production of round dowelling. The roughly-shaped material is driven through a smooth-surfaced hole (or one equipped with teeth or serrated edges) in a thick iron plate, and in this way a completely cylindrical stave with the same diameter over its whole length is quickly and easily produced.

Bolts were carried in leather quivers, but little information about them is available. In 1444 the city of Bern purchased quivers from a saddle-maker and from a shield-maker (118:45). The quivers intended for hunting were often covered with hair, as seen in figure 28.

The English scholar Payne-Gallwey tested three different crossbows of 15th-century western-European type. The largest, which he regarded, incorrectly, it would seem, as being a rampart crossbow, weighed 8.16kg and had a steel bow 97.2cm in length, 6.35cm broad and 2.5cm thick at the middle. The draw length was 17.78cm, the draw weight 1200 English pounds, which equals 544.3kg. It was spanned with an English windlass. The bolt was 35.6cm long and weighed 95g. A range of 450 yards (405m) was achieved with this crossbow (75:14-15).

The second crossbow had a steel bow 80cm long, 4.1cm broad and 1.9cm thick at the middle. The crossbow weighed 7.3kg and the draw length was about 15.24cm It had a range of between 380 to 390 yards (342-351m) (75:21).

The third and smallest crossbow that Payne-Gallwey tested was the one most similar to the

---

40 There is also evidence that arrows were still shot from cannon at this time. For instance at the Second Battle of St Albans, fought in 1461, the Burgundian mercenaries of the Earl of Warwick were using cannon 'that would shoot both pellets of lead and arrows of an ell of length with six feathers, three in the midst and three at the other end, with a great mighty head of iron at the other end, and wild fire with all'. (*Gregory's Chronicle* in J Gairdner, ed, *The Historical Collections of a Citizen of London*, London 1876, pp.211-4). Experiments with shooting arrows from both artillery and hand portable firearms have continued into the present century, and fire arrrows were certainly shot from both in European naval warface until at least the end of the 18th century.

Figure 28 Martyrdom of St Sebastian. To the right a crossbowman whose crossbow is being spanned with a cranequin. Second half of the 15th century.

typical 15th century military crossbow. It had a tiller 91cm long. The steel bow was 76.2cm long, and 4.1cm broad and 1.27cm thick at the middle. The draw length was 15.24cm but unfortunately the draw weight was not specified. It weighed 6.35kg. Shooting a bolt 31.75cm long and weighing 72g. the range achieved was 330-350 yards (297-315m) (75:22). This crossbow resembled the one illustrated in figure 20.

The effective range of the crossbow, however, was certainly much shorter than the extreme range. In the first half of the 15th century one hundred paces (75m) seems to have been a reasonable range for shooting with the crossbow (90:407). By the middle of the century and during its second half the usual range was somewhat longer, 120-125 paces or 90-94 m. (91:330).

On the basis of his own experience Payne-Gallwey considers that one can loose one shot each minute from a crossbow spanned by the English windlass, and that a little longer time is required when using one that is spanned with a cranequin (75:37). It seems likely that a considerably greater speed of shooting could be achieved with a crossbow spanned by a spanning hook or with a *krihake.*

For comparison with the figures given above, it should be noted that a modern longbow intended for use by an adult male has a draw weight of 50 or at most 60 pounds - that is to say, 22.68 or 27.2kg respectively. In France longbows with 36 or 54kg draw weights are still used from time to tlme (13:7-8). The renowned English longbows of the 14th and 15th centuries appear usually to have had a draw weight around 70 pounds (31.75kg) or a little more (13:57).[41]

The arrows for the English longbow were usually 70cm long. In trials using a close copy of an English longbow, and arrows 71cm long weighing 42g, a range of 190 yards (171m) was achieved. With arrows weighing 20.1g the range increased to 245 yards, (around 220m). Arrows that weighed 84g and that were 88.9cm long, shot with a bow that had a draw weight or 85 pounds, (38.5kg) went only 112 yards, (around 100m) (13:57). Even during the Middle Ages and in the 16th century a range of 250-260m was regarded in England as being unusual (75:21-22). However, six shots could be got off in a minute with the English longbow (75:37).

In all German cities in the 15th century target shooting both with handguns and crossbows was practised. It was treated seriously and much effort was expended on it. Some shooting was at targets, and some at popinjays. Participation in part, at least, was obligatory,

---

41 There seems to be no evidence to corroborate this statement. Research into the, admittedly later, long bows from Henry VIII's warship Mary Rose (sunk 1545), suggests that draw weights well in excess of 100lb (45kg) were not uncommon.

certainly for younger men; and they had to start at an early age. In Bern two dozen small crossbows for boys were acquired at the city's expense in 1437 (118:40). Shooting took place every Sunday and holy day. There were plenty of holy days in the Middle Ages, which explains why during 1417 in Naumburg, for instance, target shoots took place on 105 days (79:650). As I have mentioned previously, the city paid for the catering, not least the beer, at these shooting exercises. Prizes too were quite often provided at the city's expense.

At times marksmen from friendly cities were invited to take part in these competitions. In these invitations the rules for the competition which were usually very precise were also laid down. The prescribed diameter for the bolts to be allowed in the competition was indicated in the letter of invitation by a circle of the proper calibre - most frequently 1.5cm, but sometimes as much as 2cm. The bolts were scrutinized by the competition secretary or by a committee. Once the diameter of the shaft had been found to be correct, the bolts were weighed and carefully checked, after which the weight was inscribed upon the shaft (91:330). In the armoury at Skokloster, Sweden, there are a number of crossbow bolts marked on the shaft in ink '3 *lots*', [or] '4 *lots*'.[42] After the word *lot* there appear on a number of bolts some illegible figures and a symbol. These represent the weight, indicated in terms of the *lot* and *kvintin*.[43] The bolts have been badly damaged, but the weights, 42-65g, coincide closely with the weights indicated in *lots* (24:174).

For the shooting competition at Sursee in Switzerland in 1452, the range was set at 120 paces (30m).[44] According to letters of invitation from upper Bavaria in 1467 shooting was to be at a range of 260 feet and the target was to have a circle 12 inches in diameter. In a letter of invitation from Ulm dated 1468 it is stated that the circle is to be of 13cm diameter and that each marksman is to discharge 40 shots. The range is given as the length of a cord sent witn the letter. In Augsburg in 1470 the range was 125 paces (94 m), and the number of shots for each marksman was 43. At Schweinfurt in 1473 the marksmen were allowed to shoot either standing or sitting, at a range of 135 paces (102.5 m) (91:330).[45]

The crossbow was a much-used weapon in Sweden throughout the 15th century. It had by then become particularly common among the peasants in the southern and central parts of the country, and by the middle of the century it was close to becoming the national weapon. On a number of occasions, for example in 1437, 1441, 1462 etc, it was ordained that crossbows should not be carried in peacetime (49:308-309), but this prohibition had no effect at all. Any directive instructing peasants to surrender their crossbows unfailingly brought forth disturbance and uproar, no matter who might have issued the order.

However, the longbow appears to have been popular for a long time in northern Sweden and in Finland, and also in Lapland, where it was still in use at least until the second half of the 18th century (13:63). When Kastelholm in the Åland Islands was taken in 1433 handbows were used (59:35).[46] It is related that in 1460 a burgher of Arboga threatened the town bailiff with a longbow (10:122).[47] It is doubtful whether in Engelbrekt's time the men of Dalarna had

---

42 The estate of Skokloster which lies some 30 miles north west of Stockholm was given to Herman Wrangel by King Gustavus Adolphus in 1611. He created a great country house on the estate by combining two medieval buildings. His son, Carl Gustav Wrangel (1613-73), built up a great collection of arms and armour which under the terms of his will has been kept together there since his death.
43 A *lot* weighs approximately half an ounce (14.5 g.); a *kvintin* weighs approximately ¹/₅th of a lot.
44 Sursee lies at the northern end of the Sempachersee in Switzerland.
45 Schweinfurt lies on the river Main some 50 miles north west of Nuremberg.
46 The Åland Islands, now part of Finland, lie between Finland and Sweden at the southern end of the Gulf of Bothnia. Kastelholm was a castle built in the 1380s which fell into ruins in the 17th century.
47 Årboga is a town in central Sweden some 75 miles west of Stockholm.

any very great number of crossbows.[48] The 1435 seal or Dalarna shows a strung longbow and an axe.[49] The copper and iron miners surely had the best and most modern equipment and weapons that could be acquired at the time, but *'the common folk throughout all the Dales'* presumably had more bows than crossbows.

The Swedish crossbow of the 15th century was presumably generally similar to the North German type and had bows of horn or of wood. Olaus Magnus speaks of crossbows of Norwegian pine and of larch (71:10). By the middle of the 15th century crossbows with steel bows had presumably been introduced from abroad. It is conceivable, although not certain, that such bows may have been manufactured here before 1500.

In 1433 at the fortress of Tynnelsö, which belonged to the diocese of Strangnas, there were 12 *kry*-crossbows, 29 new crossbows, tnree old 'marksman' crossbows, and eight old *vippe*-crossbows (89:3-4).[50] The *kry*-crossbow was presumably one spanned by a *krihake*. The 'marksman' crossbow was probably a sort of designation of quality, and one which as we have seen had also made its appearance in Germany. The *vippe*-crossbow was spanned with the goat's-foot lever, as I have shown above.

Some of the peasants who took part in Karl Knutsson's march to Skåne in 1452 were mounted crossbowmen who could both span and shoot from horseback. In fact they were little short of being men-at-arms, that is to say professional soldiers (59:295).

From the end of the 15th century some valuable information about crossbows and their bolts and accessories comes from the Memorandum and Accounting Books of Stockholm and Arboga, as well as those of Jönköping.[51]

As I have said, crossbows with composite bows seem to have been the most usual ones in Sweden at this time. This is confirmed by evidence from the Arboga memoranda book. A complaint was brought on October 13 1477, against *verkmästare* Anders stating that he had given an unserviceable crossbow in an exchange. The court found that the crossbow was *'false'* because it (presumably the bow) was of 'yew wood' and not horn, although *'the law'* (probably the guild regulations) said it should be of horn and not wood. Anders could not deny this, and he was therefore fined (11:96).

In the Dalarnas Museum, Nordiska Museet, the Ornässtuga Museum, and the Zorn Museum at Mora there are a number of what are undoubtedly Swedish composite crossbows dating both from the second half of the 15th century and from the beginning of the 16th century (figures 29 and 30).[52] These crossbows

---

48 The region of Dalarna, 'The Dales' lies north west of Stockholm and extends to the frontier with Norway. Engelbrekt Engelbrektsson was born sometime in the 1390s, a member of a German immigrant family. He was a burgher in Västerås (a town on the northern shore of Lake Mälar near its western end) and a 'bergsman' (mountain man) part-owning iron mines and works in the mountains north of Västerås. Discontent broke out in the 1430s over high taxation caused by King Erik's long war with the Hanse, and much of this was directed against the governor, Jösse Eriksson, in Västerås castle. Engelbrekt was elected spokesman for the citizens of Dalarna and Bergslagen (the mountain district), and travelled to Denmark in 1431 or 1432 to ask the King to replace Eriksson. The King eventually refused and Sweden rebelled. In January 1435 what is known as the first Swedish parliament, meeting at Årboga, elected Engelbrekt to the Council and put him in command of the armed forces. Following a brief reconciliation with King Erik in October, rebellion broke out again and in January 1436 Engelbrekt and Karl Knuttson (later King Karl VII) were appointed joint Commanders-in-Chief. In April or May of that year Engelbrekt was murdered while travelling from Örebro to Stockholm.
49 In the 1520s a new seal was introduced for Dalarna, consisting of two crossed 'dale-bolts' (ie. bolts from Dalarna) and a crown. When provincial coats of arms were introduced at the funeral of Gustav I Vasa in 1560, the coat of arms of Dalarna was specified as two crossed dale-bolts of gold with silver heads on a gold ground with above an open crown of gold. These are still the arms of Dalarna today.
50 Strängnäs lies on the southern shore of Lake Mälar some 35 miles south west of Stockholm. The fortress of Tynnelsö rises from a small island in the lake just north east of the town.
51 Jönköping at the southern end of Lake Vättern in southern Sweden.
52 Ornässtuga is a two storey wooden hall in Ornäs in central Dalarna, with living quarters on the first floor, and stores beneath. Legend tells that the future Gustav I Vasa escaped from Danish soldiers by being lowered by the lady of the hall through a privy on the first floor.

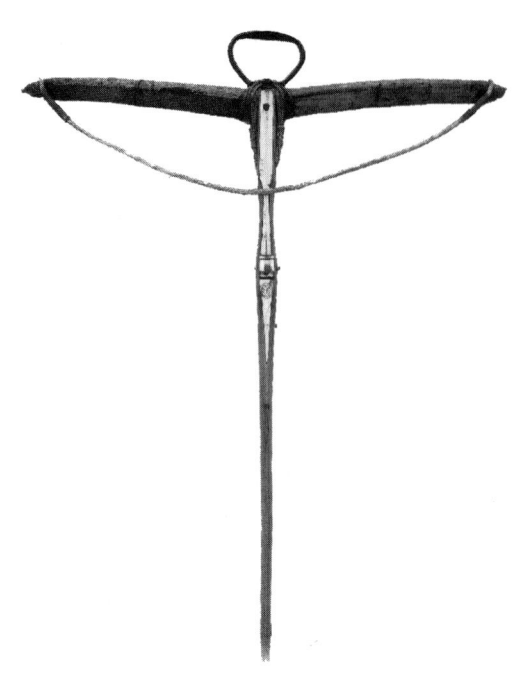

Figure 29 Crossbow with composite bow. Second half of 15th century. Swedish. String not original. Dalarnas Museet [Museum of Dalarna], Falun.

are very similar to the Austrian or South German crossbows in figures 20 and 21, but they are of an older type, larger, and with tillers of a more slender form. The lock of each has a nut of elk or stag horn, which is held in place with a binding; and a lever-trigger of iron, now lacking its spring. The front of the recess for the nut is lined with horn. A strip of horn is let into the upper side of the tiller ahead of the lock. This horn strip, and the top of the tiller, curve up towards the fore-end. A bolt-channel, narrow and shallow at its rear, has been cut into the horn strip. There are two designs for the forward part of the horn strip. In the one, which is the older, the channel becomes deeper and broader in the forward, upward-swept part (figure 29). In the other type (figure 30), another piece of horn is pinned to the top of the strip towards the fore-end, and this has a deeper and broader channel for the bolt.

In Germany this latter type appears at the middle of the 15th century. Behind the lock a horn strip of elongated triangular shape is let into the upper side of the tiller. On most tillers this strip has one or more circular patterns, and sometimes other decoration as well. At the lock the sides of the tiller are inlaid with a very slender strip of horn, tapering at each end, which runs down from the upper edge. On at least a couple of examples these strips are of dark horn, probably ox-horn. Towards the front the under side of the tiller is inlaid with a very short, heavy strip of horn. At the front this strip projects a little beyond the fore-end of the tiller, and at the back it ends in an elongated triangular point. On the under side of the tiller of some or these crossbows is a hook for the spanning-strap of the *krihake*, (figure 30). Others are meant to be used with the spanning-hook, and therefore have no hook on the tiller itself.

The crossbow illustrated in figure 29 has an overall length of 113cm. The tiller is 101cm long, and the bow 91.5cm. At the middle the bow is 6.5cm broad and 3.5cm thick. The belly consists of a number of longitudinal horn ribs, and, as usual, the back is built up with a thick layer of sinew. Around this has been bound a thin layer of sinew and on top of this again there is a leaf-thin wrapping of birch-bark, which is handsomely decorated with stencilled wedge-patterns. The bow is bound to the tiller with hemp cords and the foot-stirrup is fastened to the bow with leather straps. The crossbow weighs 3.8kg. I am grateful to the local historian, S Svärdström, for this information.

Figure 30 Tiller of a composite crossbow from the second half of the 15th century. Swedish. Nordiska Museet [Nordic Museum], Stockholm.

On a crossbow of the same type, but somewhat smaller that is preserved in the Livrustkammare, the bow is of almost rectangular cross-section, but with rounded angles and slightly bellied sides. The ends of several obliquely applied horn strips are visible at the somewhat damaged ends of the bow. The bow is clad with birch-bark, and glued over this is an outer covering of parchment decorated with very small circles, so that it looks like sharks'-skin, and stencilled with wedge-shaped designs. This crossbow was spanned with a spanning hook.

Of the crossbows just described the one in the Livrustkammare comes from Skog in Hälsingland, but the others were all found in Dalarna.[53] They are all so similar in shape and decoration that they were probably made at the same place. Most of them are very well made and all were undoubtedly made in Sweden. In the past it has been asserted that crossbows with composite bows were not made in Sweden, but were imported from Lübeck. The reason for this statement is probably the assumption that there were no crossbow makers in Sweden during the Middle Ages. However, as I shall show, at this time the craft in question was abundantly represented in Swedish cities and very probably in the countryside as well. It is also surely a little rash to assume that the crossbows described above were of German origin because they are well made, since the *verkmästare* [literally, master craftsmen] of Swedish, Danish and Norwegian cities and castles undoubtedly did just as good work as did contemporary German crossbow makers.

In Sweden in the 15th century and the first half of the 16th century, a person who is recorded by his Christian name followed by the professional title of *verkmästare* is always a crossbow maker, as I have shown elsewhere. The officer of a craft guild who at the end of the 15th century began to be called *ålderman* was earlier called *verkmästare* (61).[54] In the second half of the 15th century he was sometimes called *verkmästare* and sometimes *ålderman* (99:157). However, as in the Middle Ages officers of craft guilds were selected for periods of only one year (61:47, 68, 101, 144, 194, 225), this guild title never became attached to the Christian name. Thus, for example, Tord Shoemaker was a master craftsman in the guild of shoemakers, but despite this he was never called Tord *verkmästare*. When we are dealing, for example, with an Anders *verkmästare*, the man involved is always, as we have just said, a crossbow maker.

In Germany in the 14th century, a *werkmeister* was a person in charge of large projectile engines and of shooting engines in general (74:620, 622). It is not clear what happened after that, but there is no doubt that in the 15th century in Sweden and Denmark a *verkmästare* was a man who made crossbows. This is proved by many entries in the Stockholm and Arboga memoranda books.

Thus, for example, at Stockholm in 1462 the *verkmästare* Simon delivered a *balista* in lieu of payment of two years' tax (94:106), and in 1489 the *verkmästare* Björn delivered a crossbow as payment for land tax (98:350). A *verkmästare* Knut figures frequently in the memoranda books at Arboga. In 1471 he had to pay a fine for having sold a crossbow illegally (10:341). In 1474 he was charged with having sold an old crossbow as a new one (11:33). A committee of six people, including a *verkmästare* Jöns, inspected a crossbow in 1477, found it to be unserviceable, and ordered *verkmästare* Knut to pay a fine of six marks for his 'unworthy proceedings'. As usual, he got a reduction, and ended by paying only two marks (11:89-90). As I mentioned before, in that same year, *verkmästare* Anders had used a 'false' crossbow in a trade for which he was fined (11:96). Further evidence from the first half of the 16th

---

53 The region of Hälsingland lies on the eastern coast of Sweden north of Stockholm, at the centre of the Gulf of Bothnia.
54 In Swedish, *ålderman* means a master of a trade guild.

century helps to confirm that *verkmästare* were still manufacturing crossbows.

In a book of land holdings that was begun in Stockholm in 1420, a large number of *verkmästare* are mentioned; the first reference dates from 1438, but the *verkmästare* Mattis to which this refers apparently lived a little earlier (93:83).

In 1460, there were seven *verkmästare* in Stockholm (90), apparently organised in a guild. In connection with the distribution of the estate of the widow of *verkmästare* Karl it is recorded that the son-in-law had delivered all the tools and work-in-progress that were found in the shop of the deceased to *verkmästare* Laurens. The settlement was subscribed on 12 November 1498 with, amongst others, all *verkmästare* 'in the trade in question' acting as witnesses (99:400, 402, 406).

At Arboga, according to the city's Memorandum Books, there appear to have been two *verkmästare* and sometimes, as in 1477, three (11:89, 96). The *verkmästare* Jakob, who had previously been employed by the *verkmästare* Knut in Arboga, was working at Gävle in 1483 (97:403).

According to a list of burghers holding franchise at Jönköping, dated 1456, there was then living in the Bridge Ward of that city a *verkmästare* Haemmingh (57:1), and in the Heath Ward two *verkmästare* called Hanis, although this is likely to be a double entry for the same man, as in one place (57:4) only his name is given, but in the other (57:5) that of his bailsman is noted as well. Other *verkmästare* living in Jönköping are mentioned fairly frequently in Memorandum Books of the city dating from the second half of the 15th century.

In 1498 a *verkmästare* Rasmus is mentioned at Västerås (99:347).[55] Indeed it is likely that in the 15th century the craft was represented in all Swedish cities, and that the manufacture of crossbows was also fairly extensive in the countryside. This was demonstrably also the case during the 16th century.

At Copenhagen there is mention in 1433 of Niels Pedersen *verkmästare* at Copenhagen Castle. In 1438 the same title and position was held by Jes Pedersen, who is recorded as '*Johannes Petri, Balistarius in Castro Hafniensi*', in a Latin document written in 1446. A *verkmästare* Isak Mathiesen is named in 1466, 1474, 1479 and 1488. A *verkmästare* Oluf seems to have been in King Hans's service in 1487.[56] Finally, there is mention of a *verkmästare* Gyde in 1496 (18:68-69). At Flensborg a total of ten *verkmästare* are mentioned in the second half of the 15th and the first half of the 16th century (18:79, note 1).[57]

There were *verkmästare* in Norway as well (123:125), and a *verkmästare* Peder is mentioned in 1489 at Sarpsborg.[58]

As well as the products of the native industry there were also imports. Weapons were free of duty in Sweden at the close of the Middle Ages (99:10). Imports apparently came chiefly from Germany, but also from western Europe. In the days of Erik of Pomerania there was considerable trade between Norway and the Netherlands.[59] In 1425 a permanent break in relations began between the Hanse cities on the Baltic, especially those led by Lubeck, and the cities of the Low Countries (4:90). The Danish kings favoured the Dutch for political reasons, and in 1443 King Kristoffer established the right of Amsterdam to trade at Bergen (4:93).[60] In the

---

55 The Swedish town of Västeras lies on the northern shore of Lake Mälar towards its western end.
56 King Hans (King Johann II) ruled the combined kingdom of Denmark and Norway from 1491 to 1513. Between 1497 and 1501 he was also King of Sweden.
57 Flensborg, now Flensburg is a town on the east coast of Schleswig-Holstein, Germany, at the border with Denmark.
58 The Norwegian town of Sarpsborg lies on the eastern side of the Oslofjord.
59 Erik of Pomerania was Erik XIII of Sweden and Erik XVII of Denmark. He reigned over the kingdom of Denmark, Sweden and Norway with Queen Margaret from 1389 until 1412 and thereafter on his own until he was deposed in 1440.
60 King Kristoffer (Christopher) III ruled the combined kingdoms of Denmark, Norway and Sweden from 1440 to 1448.

1480s and 1490s Dutch vessels generally came fairly frequently to Stockholm (97:235, 98:62, 99:238).

It was presumably through trade with the Netherlands that the western European type of crossbow entered Norway, where it predominated for as long as the crossbow remained in use. This was evidently also the case in western Sweden where the crossbows used by the beginning of the 16th century were the western European type (26:3). Dutch crossbows may have come to Sweden partly via Lödöse, and partly from the Danish cities in Halland or from the Norwegian ones in Bohuslän.[61]

According to guild regulations of 1425, there would seem then to have been 15 crossbow makers in Lübeck (18:80) and many crossbows were exported from that city to Sweden.

There is no mention of spanning devices other than spanning-hooks and *krihakar* in the Memorandum Books of Stockholm, Arboga, and Jönköping.[62] Thus, for example, a spanning-hook is recorded at Stockholm in 1482 (97:388). At Arboga there is mention, amongst other things, of a hook and a spanning-belt, in 1472 (10:366), and in 1486 of a spanning-hook and a belt (11:289). At Jönköping '*en span haka*' was noted in 1479 (57:295).

Simple *krihakar*, with a hook and a pulley, which apparently date from the early 15th century, have come from finds made in medieval Swedish fortresses and cities (25:81).

At Jönköping a *krihake* is mentioned in 1459 (57:15). At Arboga a *krihake* appears for the first time in 1469 (10:317), and at Stockholm there is mention of a spanning-belt and a *krihake* in 1481 (97:295).

Krihakar were used in Denmark as well. At Odense in 1496, amongst other things that blacksmiths had to make as masterpieces, was the *krihage*, and similarly saddlers had to make a spanning-belt (18:75, note 4).

Among the *krihakar* illustrated in figure 58, the left half of the *krihake* to the left in the upper row was originally a simple *krihake* that presumably dates from the 15th century. Subsequently it was converted into a double one by crudely adding the right half (25:81-82). Like this one, the *krihake* to the right in the upper row of the figure has pulleys of horn, and may well date from the end of the Middle Ages.

In Stockholm in the 1460s a crossbow was usually valued at 12 öre or one and a half marks (94:106, 423). A *krihake* was given a value of one ore at Jönköping in 1459 (57:15).

During the 15th century in Sweden crossbow bolts were often of the international type with socketed heads 6-7cm long, having short tips of rhomboidal section. Thus, for example, in excavations in the moat of Rumlaborg Castle, bolts of this type have been found with their shafts preserved, fletched with two spirally set strips of wood.[63] The length varies between 34 and 40cm (86:49-50). Rumlaborg was destroyed in 1434, but was subsequently rebuilt, and the bolts described above may therefore date from the 15th century.

Another type of bolt, the Dale bolt, (figure 31) appears to have been very common in Sweden in the 15th century.[64] These crossbow bolts have a long, narrow head with a tang and a small tip, all of square or roughly rectangular section. The tang is usually 2cm long and is thrust into the shaft, which is bound with cord at its forward end to prevent splitting. The two flights are usually carved from the same material as the shaft. These bolts are usually about

---

61 Lödöse lies on the river Gota some 25 miles north of Gothenburg. From the 12th century it was a predecessor of Gothenburg as a major port. In 1473 Nya Lödöse was established on part of the site of modern Gothenburg. Lödöse began a steady decline thereafter, but did not lose its status as a city until 1646. The region of Bohuslan extends from Gothenburg to the Norwegian frontier. The region of Halland extends south from Gothenburg down the coast to the southernmost region of Sweden, Skåne.
62 *Krihakar* is the plural of *Krihake*.
63 Rumlaborg Castle lay at the southern end of Lake Vättern near the modern town of Huskvarna. It was built in 1390 as a royal castle with a governor ruling the area around it. It was still in use in 1611 when it was re-fortified.
64 ie, a bolt from Dalarna (the Dales).

Figure 31 A Dale bolt. Length 34.4cm., weight 35g. Livrustkammaren [Royal Armoury], Stockholm.

34.5-38.5cm long, of which 10-12cm form the exposed part of the head. The weight varies from 35 to 45g.

Dale bolts were effective weapons. The *Karl Chronicle* says of the peasants who served as mounted crossbowmen in Karl Knutsson's Skåne campaign of 1452: *Truly their Dale bolts were so sharp that they went through both horse and man* (59:295). By Karl Knutsson's order each man was to carry a crossbow and eight dozen bolts (49:635).

Olaus Magnus has the same high opinion of these bolts as the *Karl Chronicle*. He reports them as being made half of iron, half of wood, and a hand's breadth and a half long, which is too short. Magnus says that the Goths carried this sort of bolt on their campaigns by the thousands, and that they were not shot point-blank at cavalry, but instead diagonally upward, so that they struck downward from above like a hailstorm (figure 32).[65] They either struck through helmets and harness and killed or wounded the horsemen, or struck the horse in the head or back, drove it wild and rendered it uncontrollable. If the bolts missed their mark and stuck in the ground, once the wooden shaft had been trodden off the sharp tang could pierce a horse's hoof making it lame and useless (70:43-44).

The *skäkta* was a type of bolt that appeared in the 15th century, but that seems to have been more common at the beginning and in the first half of the 16th century.[66] Olaus Magnus says that it was a broad (broad-headed) bolt that was used mostly against horses (70:36). Presumably, the *skäkta* had a point of half-moon shape (figure 27, no.8).

Sometimes, as at Stockholm in 1481, there is mention of another sort of crossbow bolt called *spada*, which presumably also had a broad or very broad head (97:262).[67]

In Sweden hunting bolts were also of very varied types. Olaus Magnus says that for hunting bears and wolves broad heads as sharp as razors were used, able to pierce the animal's dense, hairy pelts. Blunt bolts were used for hunting squirrels and martens, and sharp-pointed ones for wild-fowling, to penetrate the hard mantling of feathers (70:30). Of course, Olaus Magnus was writing in the 16th century, but in this case what he says is probably also true of the 15th century.

By blunt bolts Olaus meant bolts with an expanded head which were entirely made of wood (figure 33). In the Middle Ages these bolts were called '*piston-bolts*' (100:211). Both then and later the hunting of squirrels was of very great importance. When such a bolt hit its mark the squirrel's pelt was unharmed, and that was the intention.

In the moat of Rumlaborg two complete 'piston-bolts' have been found plus two good-sized fragments. The complete ones are about 39cm long. The heads of the bolts are more cylndrical than those on 'piston-bolts' of a later date, and they have a flat, unshod striking surface (86:49).

It appears that bolt heads of bone were in use in the north throughout the Middle Ages, and, curiously, bolt heads of hard wood seem to have continued in use for a very long time in Sweden. The *härbre* [barn] of Älvrosgården in Skansen

---

65 Goths here means the inhabitants of Götaland, Southern Sweden. Olaus Magnus was himself a Goth from Östergötland.
66 Literally *skäkta* means 'swingle' or 'scutcher', a wooden instrument for beating flax.
67 *Spada* literally translated is a spade.

Figure 32 Swedish peasants doing battle with cavalry. After Olaus Magnus.

apparently dates from the first half of the 16th century, or a little earlier.[68] It was the stronghold of the house, and had obviously been subjected to attack or siege. Around most of the window openings, particularly those in the gable ends, are many marks left by bolts, the points of which, made of iron, bone and juniper wood, are still sticking fast in the walls (13:54-55).

It is impossible to be certain whether these projectiles were shot from bows or from crossbows. However, in the *Kalevala* it is expressly related that, in 'saga times', people in Finland were shooting bolts wlth wooden points from crossbows. Here we read of Joukahainen, when he was preparing his crossbow and bolts in order to shoot Väinämöinen: '*Then he made a shock of arrows, fletched them on their three sides, made the shaft of oak-wood, the points of pitchy timber*' (58:80), and further on: '*Next he hardens the bolts, makes their points good and sharp, converts them into serpents with black venom made from the poison of vipers*' (58:81). Here we are evidently dealing with poisoned arrows with tips of wood.

During the troubled times of the middle and second half of the 15th century the city of Stockholm had to remain under arms, and because of this the city purchased, among other things, great quantities of arrows, much evidence for which comes from the city's Scot Books (tax registers). The iron needed for the manufacture of arrowheads was usually purchased by the cask. It is not clear how much iron a cask contained during the Middle Ages, but it appears usually to have been 25 *lis* pounds net (33:426, 427). If one then calculates, using the Stockholm standard in which a lis pound was 6.866kg (53:21-25), a cask should contain 171.65kg net. However it would, probably be unwise to regard this evidence as entirely reliable. Purchases of one or more casks of iron for arrows are recorded, for instance, in 1463 (94:380), 1464 (94:387) and 1470 (94:431). In this last year the smith Olof received four marks for forging a cask of iron into arrows (94:431).

The usual term for unmounted arrow-points was *slagen pil* [literally 'struck arrows'] (94:393) or *Ostickad pil* [literally 'unstuck arrows'] (95:80). It appears as though the word *pil* originally meant merely the arrowhead.[69] In 1465 Hinrek the Jutlander received four marks

---

68 The Swedish word *härbre* may be translated as barn or store house. Such buildings were sometimes also designed as the strong point on the farm. Älvrosgarden is a farm from Älvros in Härjedalen on the border with Norway. It is now in Skansen, Stockholm, the first open air museum in the world, opened in 1891 thanks to the pioneering vision of Artur Hazelius.
69 the term *pil* is more commonly used to describe the whole arrow or bolt.

Figure 33 Blunt-headed bolt, *bultekolv*, [piston -bolt], for squirrel hunting. Length 40.5cm., weight 110g.

for 'struck arrows' (94:393).

Arrow-shafts, *piltenar*, were frequently purchased by the city of Stockholm, as for example, twice in 1465 (94:210, 393). As I have mentioned before, ready-mounted arrows were called 'stuck arrows' (96:79). In 1440, 7$^{1}/_{3}$ öre was paid for '*30 dozen arrows to be stuck*' (94:381). In 1465, Jon the Painter was paid for arrow-shafts and for having 'stuck' struck arrows (94:393). Per the Miner appears in the records both in 1466 (94:401) and in 1469 (94:425) as a supplier of arrow-shafts, and as an 'arrow-sticker'. Sensibly, he even settled his taxes with arrow-shafts (94:111, 147, 247). In 1493 in Stockholm there is mention for the first time, of a man who was an arrow-mounter by profession, one Anders Pilstickare (99:69). In 1499 there were ten tuns of *styckat piill* ['stuck arrows'] at Nyslott in Finland (14:83).[70]

Arrows were carried in arrow-quivers, but such quivers are very rarely mentioned. A crossbow quiver with belt was valued at 12 öre at Arboga in 1470 (10:327). In 1499 at Nyslott there were two *piilekoffwer iernslaget* [iron-mounted arrow-quivers] (14:83). Presumably the quiver and accompanying bolts, together with the spanning apparatus, made up the *skyttereda* [marksman's kit]. The phrase '*A crossbow with its complete marksman's kit*' is commonly found in the Memorandum Books (97:331, 390, etc.).

Olaus Magnus reports that in Sweden during his life-time, the first quarter of the 16th century, shooting at a mark or a popinjay was practised both in cities and in the country (71:133). There is every likelihood that this was also the case in the 15th century. By that time shooting at the popinjay had become a fashionable sport. An ordinance issued on 29 April 1489 by both the bailiff of the Castle and the burghermaster of Stockholm and his council, laid down that only the guildmasters in the distinguished Guilds of Our Lady and St Gertrude had the right to set up popinjays. Shooting was to take place on the second day of Pentecost, Whit Monday. Only the members of the two exclusive guilds who were merchants might participate in the *avskjutning* [shoot]. Things proved expensive for the successful burgher or burgher's son who succeeded in shooting down the popinjay, for he had to stand the cost of a cartload of ale, which was to be consumed by the brothers and sisters of the guild at the feast that ended the shoot. A cartload of ale was 12 tons (53:36). It was a consolation for the prize-winner that under penalty of a fine of 12 marks, he was prohibited from offering anything but ale. What he got as a prize is not stated.

'*But should any journeyman* [craftsman] *venture to shoot down the popinjay, whether by stealth or openly, he shall have it set up again within twenty-four hours at his own expense immediately after it has been shot down, and on the next day of session he shall without grace pay a fine of three marks at the City Hall*' (98:340).

The oldest Danish popinjay guild is the *Hellig Legems Lag* [The Company of the Body of Christ] at Ålborg, whose Charter dates from 1441 (119:18).

Popinjay shooting was introduced at Bergen by the Germans. In 1497, the German journeymen received from the city's bishop ground upon which they might set up their popinjay (119:18).

---

70 Nyslott is the Swedish name for the Finnish town of Savonlinna, which lies in the middle of the Saimaa Lake system, in the province of Mikkelin lääni in the south east of Finland. The area was of strategic importance in the 15th and 16th centuries as it bordered on Russia. The town grew up near the Castle of Olavinlinna (in Swedish Olofsborg) which was built between 1475 and 1477.

5 The Crossbow after 1500 Central Europe

In the 16th century the crossbow reached the pinnacle of its development through the need to compete with hunting and target guns and by borrowing certain features from them.

The old crossbow release mechanism, consisting simply of a nut and a trigger-lever, and, occasionally a lever spring, had a very hard pull, particularly when there was a considerable distance between the forward end of the trigger serving as a lever sear and the bolt upon which the trigger pushed. At the beginning of the 16th century in Germany the crossbow was improved by the introduction of a special sear-lever between the trigger and the nut. This piece, 2 in figure 34, pivots on a pin. At its upward end it has a tooth on its front side. A piece of strong cord, 4, runs through a hole in the lower end of the sear and through a hole in the lower side of the tiller. At each end it has a permanent knot, the upper one preventing it from being pulled in through the hole in the tiller. The forward part of the trigger, which has a tooth on its upper side just behind the end, is pressed upward by the lever spring 5. When the

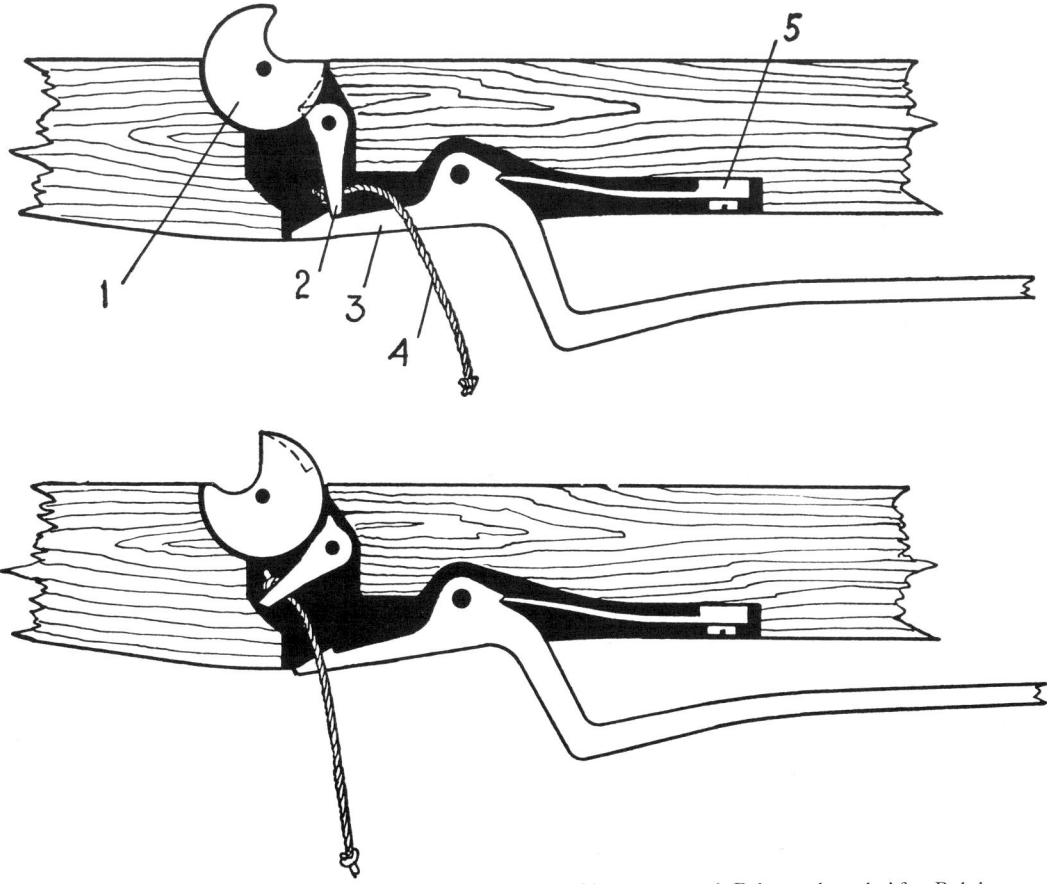

Figure 34 Crossbow with a sear. First half of the 16th century. Above, spanned. Below, released. After Rohde.

crossbow has been spanned and the nut has come into the correct position, the cord is pulled, whereupon the lower end of the release-piece is drawn backward and engages behind the tooth on the upper side of the trigger just described. At the same time, the tooth at the upper end of the sear's side engages in the spanning notch of the nut and so holds the nut in the spanned position. Now only a slight pressure on the trigger is needed for the shot to be discharged. For this reason a spanned crossbow of this type was decidedly dangerous (81:100). Because of this, in about 1500 the Emperor Maximilian had this release mechanism installed on his crossbows together with a small pivoted arm fastened on the lower side of the tiller. This could be dropped into a vertical position to prevent the trigger from being drawn upward against the tiller. On the other hand, when it was brought up along the under side of the tiller, the trigger was able to move freely. A spring attached to the lower side of the tiller, pressed against the upper branch of this arm and held it in its two positions (19:409).

Until well into the first half of the 16th century German crossbows often had both bows and trigger springs made of horn, and yet these bows were so powerful that as a rule they had to be spanned by a cranequin (81:101).

The German crossbow illustrated in figure 35 dates from the beginning of the 16th century, has a lock as in figure 34 and a tiller 66cm long, and is intended to be spanned with a cranequin. The horn bow is composed of three almost equally thick layers, the belly consisting of a strip of wood, the core presumably of a layer of horn strips and the back of a layer of sinew. The outer envelope consists of parchment, decorated with quite large, closely spaced, circular dots. The bow is 70cm long, 4.6cm broad and 4.4cm thick at the middle. At the fore-end, the upper side of the tiller has a *pilstöd* [bolt rest], adjustable sideways, of approximately the same shape as the fixed one on the crossbow illustrated in figure 21. At a later date the crossbow has been equipped with a

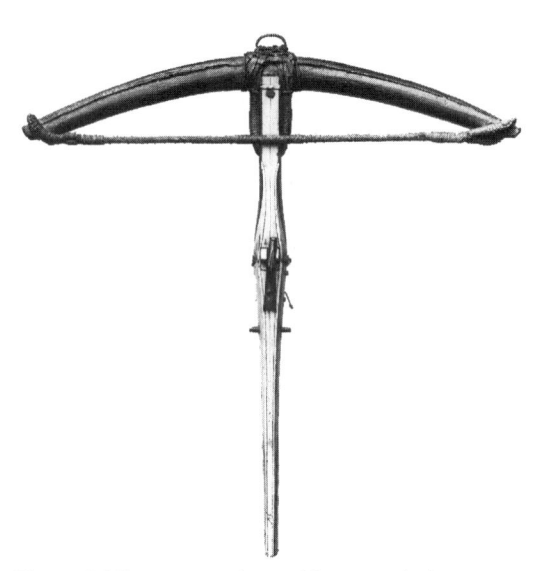

Figure 35 German crossbow with composite bow, spanned with cranequin. Beginning of 16th century. Livrustkammaren [Royal Armoury], Stockholm.

horn spring for securing the bolt.

In the middle of the 16th century, when the demand for crossbows fell dramatically, many crossbow makers had to find a new trade. They were used to working with horn and to making glue for themselves. In Lübeck at this time the crossbow-makers began to make glue for sale, at first only as a side-line, but soon as their main trade. Because of this, for many years the word *Armbrustirer* or crossbow-maker also meant a manufacturer of glue (18:82-83).

The crossbows in figures 39 and 40, manufactured at Nuremberg and Augsburg after the middle of the 16th century, are of a particularly complex design. The locks are a development of the type with the sear that has been described above. The old lever-trigger has evolved into a fixed guard which protects the trigger. The latter has been divided, so to speak, into three parts, a pressure-bar (or sear), an intermediary lever, and the trigger proper. The design is shown in figure 36. A little behind the nut there is a vertical hole g, which leads down to the intermediary lever E, which is forced upward by the spring G. The rear end of the intermediary lever is held by a notch on the higher and broader part of the trigger, F, which itself is

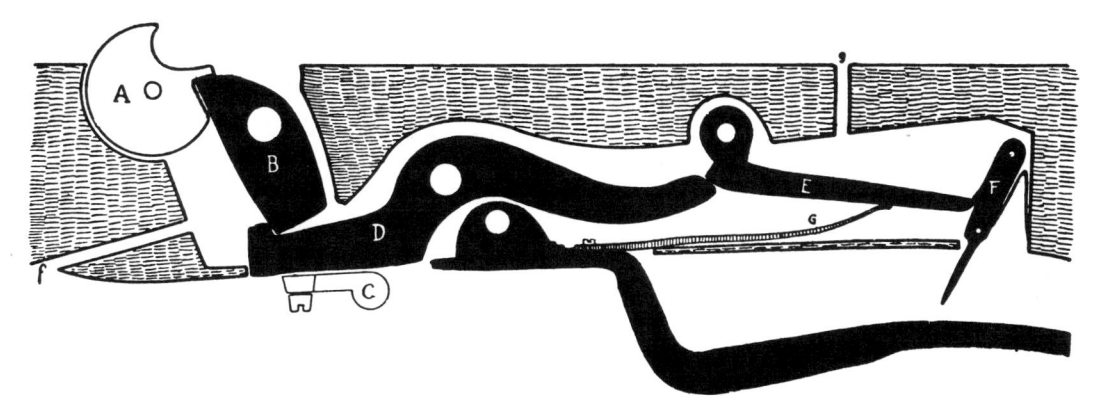

Figure 36 German crossbow lock. Second half of 16th century, or 17th century. After Wegeli.

pressed forward by a spring. This trigger has a narrow lower part and pivots around a pin, which is passed through it well up on its thicker section. On the under side of the tiller, a little ahead of the lock, another hole runs obliquely backward and upward towards the release-piece B. The pressure-bar D has a lug at the forward end of its upper side and can be locked securely by the safety key C, which rotates on a screw set in the underside of the tiller alongside the recess through which the pressure-bar projects. For spanning, the nut had to be in the cocked position and the bowstring was then drawn over it.

First it was necessary to insert a steel pin with a shaft, the *stickar* [pricker] into the hole *f* and press the release-piece B back, so that the lower end of the forward edge of the release-piece fitted behind the lug on the pressure-bar D. Next, the safety-key C was turned so that its wing lay across the tiller and the underside of the pressure-bar. After this the pricker was inserted through hole *g*, to push the rear end of the intermediary lever E into the notch on the trigger F. The weapon was now spanned with the safety catch on. The safety catch was effective, because when it was in the 'on' position the nut could not be released from the cocked position. If the safety catch was turned so that its wing lay along the tiller, the bow was ready to shoot. Only a light pressure on the trigger F was needed for the shot to be discharged.

Despite the fact that the design described above had a very light trigger-pull, the trigger F was soon replaced on a number of crossbows, particularly those that were intended for target shooting, by decidedly complicated set-triggers (figure 37). These set-triggers were usually mounted on an iron plate let into the tiller. A typical set-trigger consists of a trigger, 5, an angled spring, 2, an arm, 1, a bar-spring, 3, and a bar, 4. The powerful upper limb of the spring, 2, presses the back of the arm upward while the lower limb presses against the rearward extension of the trigger. The bar-spring, 3, presses down on the long backward extension of the bar, 4. A short cord is attached to the rear end of the arm, 1, and passes out through a hole in the underside of the tiller.

When the bow was spanned, this cord could be pulled, drawing down the back of the arm,

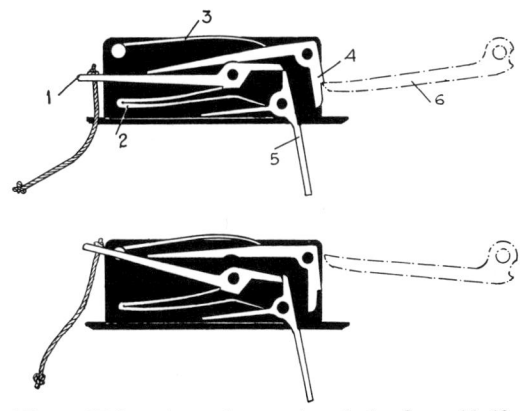

Figure 37 Set-triggers for crossbow locks. Second half of the 16th century, and 17th century. After Rohde.

57

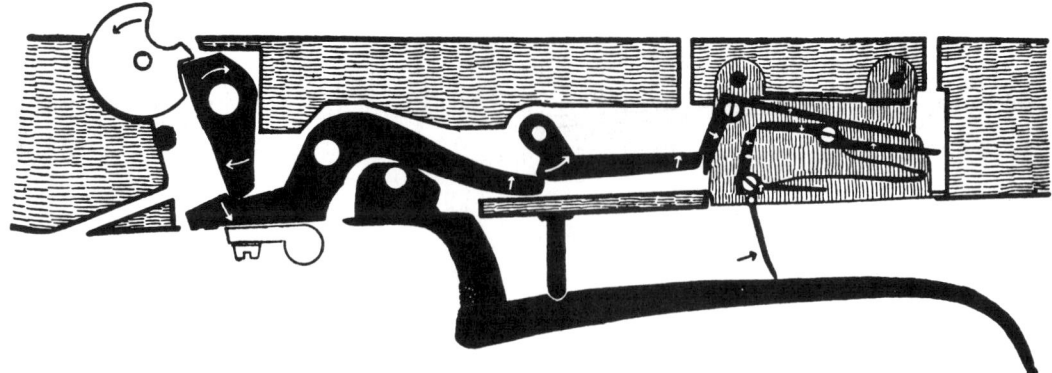

Figure 38 Crossbow lock with set-trigger. Second half of the 16th century, or 17th century. After Wegeli.

1, and engaging the front of that arm on the upper end of the trigger, 5. Next, a pricker could be used to press down the intermediary lever, 6, so that its rear end fitted into the notch on the bar, 4.[71] Now only a very light touch on the trigger was needed to move its upper end forward and release the front of the arm, 1, the other end of which was thrust rapidly and with force against the rear portion of the bar, 4, by the spring, 2. This bar, 4, released the intermediary piece, 6, and the shot was discharged. On these locks the safety-device was of the same sort as on the type of lock previously described.

Often, the arm did not have a cord through its rearward end, but instead a hole was drilled through from the top of the tiller, into which the pricker could be inserted in order to press down the rear end of the arm, 1 (figure 38), and thus cock the hair-trigger.

Such weapons (figures 39 and 40), are usually quite small but they are very strongly built. They have very powerful steel bows, usually between 60 and 70cm in length. Tillers range in length between 55 and 70cm, but the majority are a little over 60cm long. The nut is stouter than on the older types. Behind the lock the tiller usually tapers slightly in section from bottom to top, and there is a suggestion of a cheek-piece. Often, on the top of this part of the tiller is a diagonal recess for the thumb. Forward of the lock the upper side of the tiller is overlaid with a flat horn strip in one piece, which has no bolt-groove. About 1cm behind the fore-end the top of the tiller is cut with a transverse dovetail slot, into which is fitted a bolt-rest in the form of a piece of bone or horn, rectangular and usually about 0.7cm thick, 1.5-2cm broad and 1-1.5cm high. This bolt-rest projects about 0.7-0.8cm above the top of the tiller, and its upper side is notched to retain the bolt. It was usual for the rear of the bolt to be fitted between the fingers of the nut, and the

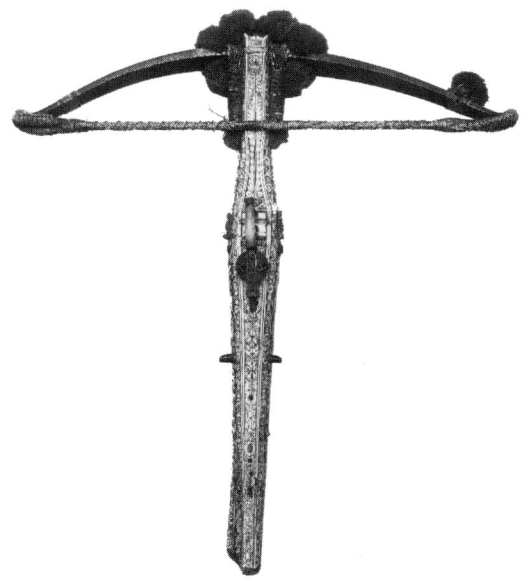

Figure 39 German crossbow with steel bow, spanned by a cranequin. About 1600. Livrustkammaren [Royal Armoury], Stockholm.

---

71 Alm actually wrote that the end of this intermediary lever fitted into the notch on the trigger, but this is undoubtedly a mistake. It must fit the notch on bar 4. This is the only correction made to Alm's text itself.

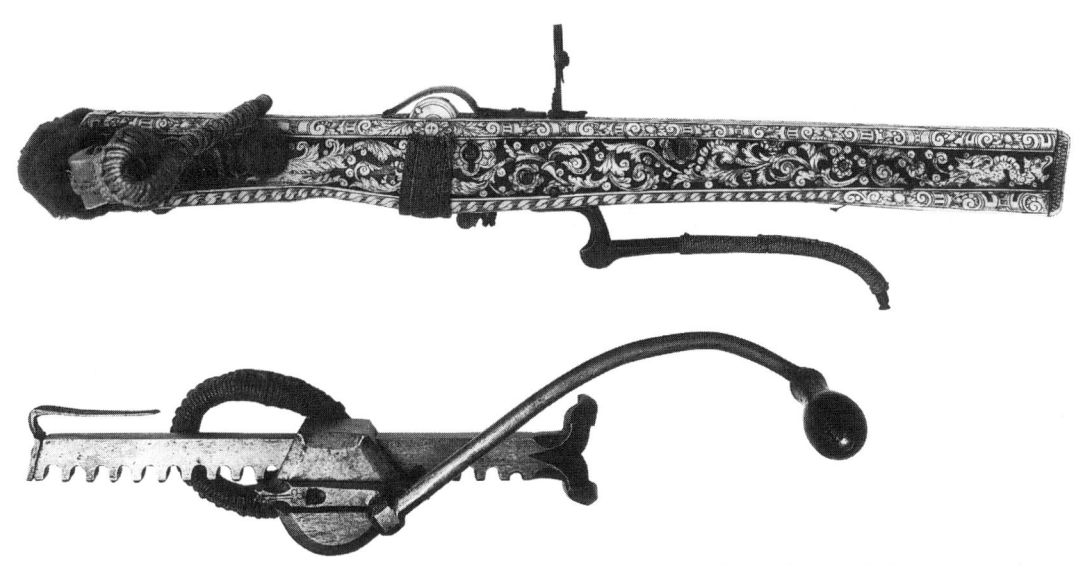

Figure 40 German crossbow with steel bow. Around 1600. The bolt-clip is pivoted forward. Below, a cranequin. Livrustkammaren [Royal Armoury], Stockholm.

front of the shaft to lie in the rounded notch of the bolt-rest. In this way, no appreciable friction arose when the bolt was discharged. These crossbows also have a spring of horn or steel (figure 40). The rear of the bolt shaft was held between the forward end of this spring and the tiller, thus keeping it still during aiming. This bolt-retaining spring could be swung back when the bow was being spanned.

The tiller was usually decorated with inlays of staghorn, the inlays themselves often richly decorated. Sometimes the whole tiller was covered with horn or ivory, embellished with incised or carved ornament and figures. In addition, these crossbows have sights, usually peep-sights, and these were often of very complicated design. Like the lower part of the trigger, the sight could be hinged forward so as not to be in the way while spanning. The bow was secured to the tiller by cord binding. Investigation of these bindings showed that the linen threads were first saturated with marrow [neat's foot] oil and were then waxed and loosely twisted together into a heavy cord which was used for the binding. The back of the bow is fixed with a semi-cylindrical piece of wood as wide as the bow, but somewhat longer than the width of the fore-end of the tiller. The binding, which is often painted gold, passes over this piece of wood. The iron ring, by which the crossbow was hung up on a wall or such like, is let in behind this piece of wood

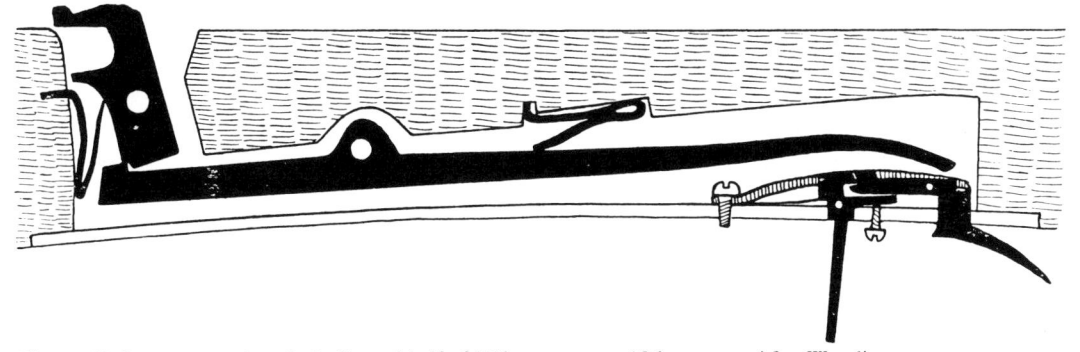

Figure 41 German crossbow lock. Second half of 17th century, or 18th century. After Wegeli.

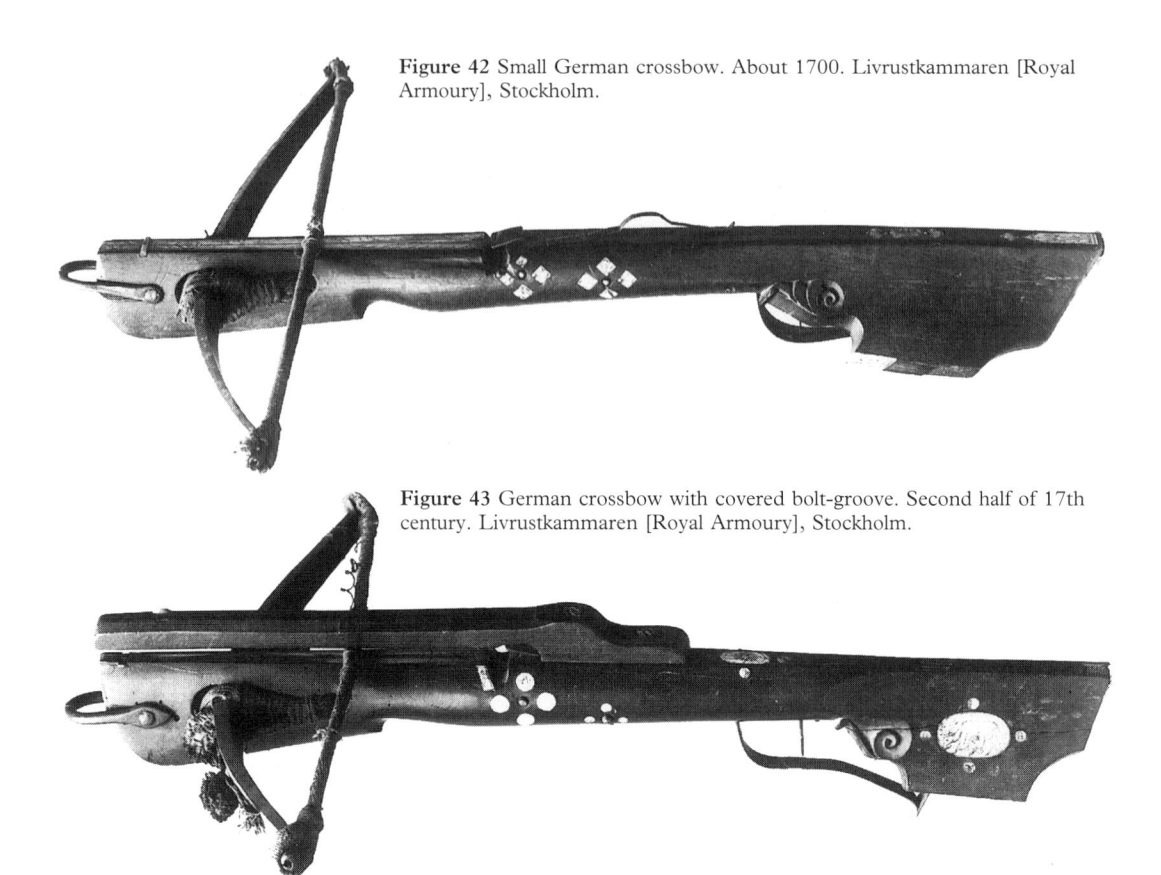

Figure 42 Small German crossbow. About 1700. Livrustkammaren [Royal Armoury], Stockholm.

Figure 43 German crossbow with covered bolt-groove. Second half of 17th century. Livrustkammaren [Royal Armoury], Stockholm.

(35:241). The forward side of the bow and the lashing are often decorated with pompons of woollen yarn, which are held in place by cords.

One large crossbow of the type described above had a tiller 69.5cm long, and a steel bow 78.7cm long, 5.2cm broad at the middle, and 1.5cm thick. It weighed 6.35kg. The spanning gap was 15cm and the draw weight 495.45kg (35:241-245).

When shooting with a crossbow of this type, the rear of the tiller was placed against the cheek, just like the butt of a wheellock hunting rifle. In Germany, in the middle of the 16th century, light forks were sometimes used to support the crossbow when shooting. When the fork was not being used it was carried on a loop slung round the neck (64:88 note).

The hunting bolts used with these crossbows were of the same type as those used during the 15th century. In the middle of the 16th century, Count Georg Ernst of Henneberg sent Duke Albrecht of Prussia a present of a crossbow with a cranequin, a quiver with bolts for big-game hunting, and a box full of *mejslar* [literally, chisels], bolts with chisel-shaped points.[72] These latter bolts were intended for hunting cranes, geese, bustards, swans, ducks, blackcock and capercaillie, as well as roe deer. In an emergency they could also be used against big game (64:88, note).

Crossbows of the type described above were used in Germany for hunting until well into the 1600s, and in the Tyrol and Switzerland for hunting chamois as late as the beginning of the 18th century (75:205). During the 17th century, weapons of this type were commonly

---

[72] Henneberg lies in the south of eastern Germany, just under 20 miles south west of Sühl, which was once part of the county of Henneberg. Albrecht of Brandenburg, Grand Master of the Order of the Teutonic Knights, seized its territories in 1525, at the same time renouncing Roman Catholicism and embracing Lutheranism. He was acknowledged as Duke of East Prussia, holding the title from the King of Poland.

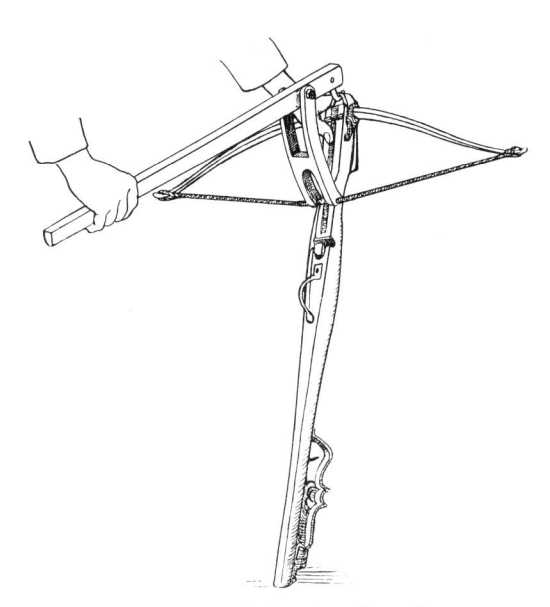

Figure 44 How the goat's foot lever is used.

used by poachers throughout Austria and Germany, and in Lower Austria they were still being so employed in the first half of the 19th century (22:317).

Smaller crossbows, for hunting small game and for target shooting, were very common in the 17th and 18th centuries. They usually had locks that were different from those of the older crossbows. Such a lock is illustrated in figure 41. It has evolved from locks of the type illustrated in figure 36, the nut having developed into a claw which retains the spring. The set-trigger mechanism is of the usual type found on hunting rifles, with a forward and a rear trigger of different shapes. The forward one is the real, or set, trigger, and the rear one may be called the 'setting trigger'. The powerful setting spring has a projection at the left side of its rear end, which lies upon and presses down on the setting trigger's upper arm just behind the pin on which the setting trigger pivots. To use the set-trigger the setting trigger must be pulled back until a tooth at the end of its upper arm slips into a notch on the setting trigger. A touch on the needle-like set-trigger will release the setting trigger, the forward horizontal arm of which is quickly and forcefully thrust against the rear of the sear. This type of lock is found in a number of variants.

Crossbows with locks of this sort were made in a number of different forms. A common type produced in the middle and latter half of the 16th century and for a great part of the 17th century (figure 42), has a lock of the design illustrated in figure 41. Forward of the lock the top of the tiller has a short bolt-groove, and there is also a bolt-rest at the fore-end. These crossbows always have a spring to retain the bolt. On the left side of the rear of the tiller is a cheek-piece which extends quite far downward. The trigger-guard is long and is fitted with finger-rests. There is usually a fold-down back-sight, and sometimes the socket of the bolt-head has a pin to act as a fore-sight.

In the 17th century the tillers were straight, but in the 18th century they often curved down to the rear of the lock. As with the wheellock rifle, the tillers of these crossbows were placed against the cheek for shooting.

Another type of crossbow much favoured at this time resembled the one described above, except that the bolt groove was covered so the string moved in a slot between the cover and the top of the tiller. These weapons usually had butts of wheellock rifle shape (figure 43). At the end of the 17th century they were sometimes provided with shoulder stocks of the type we still regard as modern and which were very common in the first half and the middle of the 18th century. In all these crossbows the bow was held on with bindings, and at its fore-end the tiller had a ring or a stirrup.

Crossbows of all the three types described above were spanned with a wooden bending lever (figure 44), which was presumably developed from the old goat's foot lever. It consists of a bar and an arm, and operates by pushing the string into the spanned position, as illustrated in figure 44. It was possible to span very powerful crossbows with this type of bending lever.

In the 18th century, the sport of crossbow-shooting continued, especially at the Saxon

Court, using both crossbows of the types described above and weapons of the 16th century type. For popinjay shooting or bird shooting (*Vogelschiessen*) as it was usually called there, bolts with a *Krönung*, a crown or quarrel head, were used. These had an iron head in the shape of a very broad, squat, four-spiked crown from the flat base of which emerged a tang which was driven into the shaft, reinforced with an iron collar at its fore-end. Such quarrels are very heavy. Curiously, it is this type of quarrel which is sometimes stated in reference books to have been designed to penetrate armour. Sharp-pointed bolts were used for shooting at targets that were divided into twelve squares. Bolts and shooting accessories were kept in small boxes or chests specially made for the purpose, often lavishly fitted, which in addition to the things mentioned above, also frequently contained a telescope (30:56), which was presumably used for marking scores.

Even during the 19th century there was a certain amount of target shooting with the crossbow. Shooting at the popinjay continued in some areas, especially in Switzerland and southern and western Germany.

The Swiss crossbows seem in general to have had shoulder stocks, butts and locks as in figure 41. They could be very large. Indeed, one dated 1897, with a shoulder stock, that I have had the opportunity to examine had a tiller 84cm long. The steel bow, 84cm long, was 4cm broad at the middle and 2cm thick, and it was wound with green bands. It was held in place by means of an iron bridle. The weapon weighed 7.45kg and was spanned by a 'pusher' bending lever. Similar crossbows are still in use in Switzerland today.

In Swiss schools, crossbows are used to teach young people the first principles of shooting. The standard crossbow used for this purpose is no toy. The tiller has a shoulder stock and is 70cm long. The lock is reminiscent of that of an Italian stone crossbow, except that it has a claw release and a set-trigger. There is a back-sight with a semi-circular notch to the right of the lock, and a fore-sight on the right side of the iron bridle attached at the fore-end of the tiller to hold the bow fast. This bridle is held in place by means of four strong screws let in from the front. The steel bow is 60cm long, 4cm broad and 0.8cm thick at the middle. The weapon weighs about 3kg. It is intended for use with a 'pusher' bending lever. The bolts are oddly fashioned with three flights carved out of the same piece of wood as the shaft.

A number of different types of crossbows, ranging from very powerful ones to toys, were manufactured in Germany until 1939, and they were listed in the catalogues of the large wholesale sporting-goods firms. A type with a straight tiller, which was very popular in Germany, was almost of 17th century type. The tiller was 85cm long and the lock was like the one illustrated in figure 41, with a set trigger. There was no bolt guide, but there was a transverse bolt-rest on the top of the fore-end of the tiller. The crossbow had a pivoting steel bolt clip. The steel bow was 63cm long and was held in place by bindings. The bow was spanned with the 'pusher' type of bending lever. Normally such crossbows weighed between 6 and 8kg. The bolt was 27cm long and the one equipped with the 'quarrel head', which was used for shooting birds, weighed 100g. The price varied depending on the fittings, between 175 and 225 marks.

Crossbow shooting was particularly popular in the city of Dresden, where there were two clubs devoted to this form of sport. The larger and more distinguished one, *die Priwilegierte Bogen-Schützen-Gesellschaft*, had about 400

Figure 45 A bolt with a crown-shaped head. Length 31.6cm., weight 77g. Livrustkammaren [Royal Armoury], Stockholm.

members at the beginning of the 1900s. Bird-shooting on the *Vogelwiese* ['birding field'] was one of the great events of the year, at least until 1918. The bird was 13ft high and 8ft across its outspread wings. It was constructed of many small segments and was literally shot to pieces in the same way as the bird used for bird-shooting with rifles. There were various prizes for shooting down the different pieces, but the person who shot down the final part became the shooting-king (75:234f). The bird was fixed to a pole 42m high, and the shooter stood 30m away. There was a special rest on which to support the heavy crossbow while shooting (79:661, note 27).

The crossbows used for this competition had tillers 80-90cm long. The bow was made of the best spring steel, 90cm long, 5.5cm broad at the middle and 1.75cm thick. At the ends the breadth was 4cm and the thickness 1.25cm. The string consisted of 25 turns of thread.[73] In all, the weapon weighed about 10kg. The bolts were 35-40cm long and weighed 175-220g each. Their heads consisted of a four-spiked 'crown', and the rear end of the shaft was shod with a horn strip. There were no flights. In shooting, the 'head' was set so that one of its spikes was uppermost and could serve as a fore-sight. The crossbows were spanned with cranequins (79:661, note 27), and were of the strongest sort, with a draw weight exceeding 500kg. Records of bird-shooting in other parts of Germany show that the marksmen had their initials engraved into the striking surfaces of the 'head' which were thus stamped into the part of the wooden bird that the bolt hit and shot down. In this way there could be no doubt as to who had shot down any particular piece.

At the great shooting competition on the Dresden Birding Field it was customary until 1918 for the King of Saxony or some member of the Saxon royal house to shoot the first bolts. The crossbows that the members of the royal family always used were made between 1741 and 1747 by J G Haenisch of Dresden (30:33), although apparently only the steel bows and the cranequins dated from this time.[74] Until as late as 1918, the Saxon royal family employed a special court crossbow 'spanner' whose duty it was to keep the royal crossbows in good repair and working order. The last of these also belonged to the Haenisch family mentioned above, members of which worked for the Saxon royal family as crossbow makers or 'spanners' for many generations (82:56).

Trap crossbows were used in Germany and Austria in the 17th and 18th centuries. Berger illustrates one for catching lynxes that is very similar to the ones used in Småland. As with the Småland ones, the shooting mechanism is on the right side of the tiller. The bolts are very distinctive, having a trident head like that of a fishing spear (16:331).

---

73 25 threads turned together do not seem sufficient to make a bow string. Alm perhaps meant 25 strands turned together, each strand already being a multiple of a number of threads.
74 Johann Gottfried Haenisch the elder (1696-1778) made and refurbished crossbows for the Saxon court for many years before 1741, and probably after 1747 as well. It has so far proved difficult to distinguish his work from that of his son, Johann Gottfried Haenisch the younger (1728-57).

6 The Crossbow after 1500
Western Europe and Spain

In France and Spain at the beginning of the 16th century, the crossbow was still to some extent in use as a military weapon. At the battle of Marignano in 1515 two hundred mounted crossbowmen formed part of Francis I's bodyguard.[75] It was about 1518-20 that the crossbow disappeared from the French armies (75:48).[76]

A company of crossbowmen formed part of the army with which Cortez conquered Mexico in 1521. When Pizarro sailed from Panama in 1524 to explore the coast of South America, his force consisted exclusively of crossbowmen, although when he conquered Peru between 1532 and 1534 his army included only a dozen men who carried crossbows (75:48).

On de Soto's expedition in the Mississippi area in 1538-42, half of the infantry were armed with crossbows and the other half with arquebuses. De Soto claimed that in a battle fought against the Indians in 1539 the natives shot three or four arrows in the same time that his crossbowmen discharged one bolt (121:435). It is interesting to note that only a few decades ago an American scholar found some old European crossbows among the cult objects of an Indian tribe in North Carolina (121:435-436).

In western and southern Europe, the crossbow continued in use as a hunting weapon until around the middle of the 17th century. Some French writers on hunting, notably Robert de Salnove in *La Venerie Royale*, published in 1645, reported that as late as the reign of Louis XIII (1610-43) crossbows were used for stag-hunting in France (75:51).

Working at Lisieux in Normandy in the 16th century was a family of outstanding locksmiths, clock-makers, crossbow-makers and gunsmiths - the Le Bourgeoys. One or the family, Marin Le Bourgeoys, probably invented the flintlock during the first decade of the 17th century.[77] In 1605 he received travelling expenses in order to deliver to Henri IV a gun, a hunting horn and a crossbow, all of his own manufacture (66:34). Interestingly, when Master Marin put his flintlock together he borrowed a number of details, from the locks of crossbows.[78]

As late as the 1590s crossbows appear to have been used for stag-hunting in England, at least by professional huntsmen (75:51, figure 20).[79]

From Spain very interesting information about crossbows and hunting with them comes from Alonso Martínez de Espinar's *Arte de Ballestería y Montería*, Madrid, 1644. This author makes great complaint that guns were replacing crossbows as hunting weapons. He describes a weapon that is dying out and a disappearing form of hunting.

According to Martínez de Espinar's description, Spanish crossbows in the middle of the 17th century were not much different from those used in western Europe in the late Middle Ages, and were thus very old-fashioned compared with contemporary German crossbows. He considered the best pattern of tiller to be one which was straight from one end to the

---
[75] Francis I ruled France from 1515 to 1547.
[76] The use of crossbows in the French army was not officially ended until an ordinance of Charles IX in 1567.
[77] More recent research has questioned this story, first told by Torsten Lenk, and has suggested that what we today think of as the true flintlock was more probably invented in France in the early 1620s. (See, for instance, W B Gusler, J D Lavin, *Decorated Firearms 1540-1870 from the Collection of Clay P Bedford*, Williamsburg 1977, pp.2-6.)
[78] Henri IV ruled France from 1589-1610.
[79] At this time crossbows were still being used for deer hunting by Queen Elizabeth and her court (see H L Blackmore, *Hunting Weapons*, London 1971, p.201).

other. Such a tiller had a bolt-groove of horn, which was swept upward a little in front so as to diminish friction between bolt and tiller. The nut was made of staghorn, as usual, with the groove for the string reinforced by a small inlaid piece of steel. The recess for the nut was also of staghorn. The lock had a long lever-trigger. There is no mention of a trigger-spring. The tiller had iron reinforcements to either side of the lock, and these plates were let in flush with the surface of the wood. Iron bridles held the steel bow, and ahead of the centre of the bow was a small iron ring for hanging up the crossbow. The part of the tiller backward of the lock del Espinar called the grip or the tiller shaft. The tiller he said, should weigh as much as the bow or else the crossbow would kick (75:149-152). Martínez de Espinar stated that the royal crossbow-maker Juan de Lastra was the only outstanding master of his craft in Spain in 1644 (75:149).[80]

These crossbows did not have sights; instead aiming was done in the old way, over the right thumb and the head of the bolt (75:152). They were spanned with cranequins (75:148).

Martínez de Espinar described several different types of bolts: one sort was large and heavy and had no very great range. Such bolts were used for hunting hares and rabbits at night, when they were easy to hit by moonlight. Bolts for partridges were a hand's breadth longer than ordinary bolts, and had heads in the form of an iron button. He also mentioned bolts with chisel-shaped heads, but did not say for what they were used. In addition he described long, and presumably also light, bolts that had very sharp points and a transverse pin midway along the shaft so they should not pass right through the prey. These were used for hunting rabbits, the long shafts being to prevent a wounded animal from crawling into its hole (75:154).

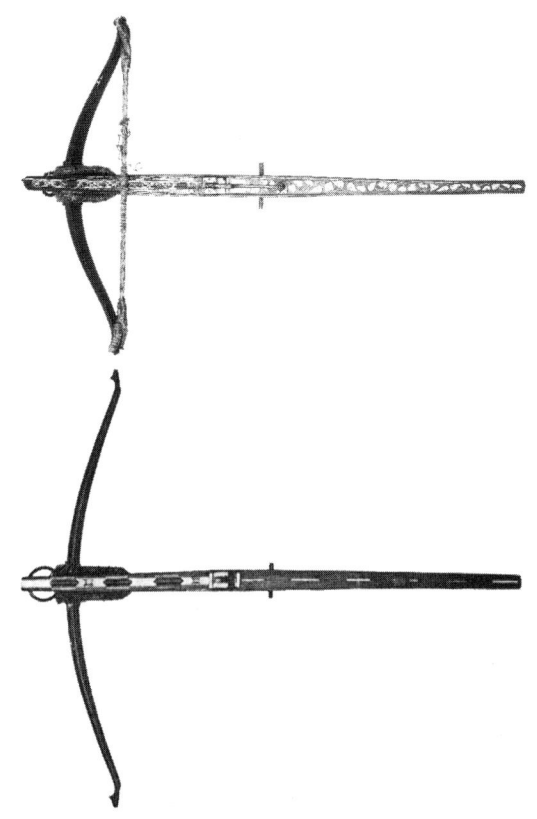

Figure 46 Small Dutch crossbows with steel bows, spanned by goat's-foot levers. Top: about 1600, Jean Jansson collection; bottom: from the second half of the 16th century, Livrustkammaren [Royal Armoury], Stockholm.

The most remarkable thing, however, is that Martínez de Espinar described poisoned bolts, used for hunting stags, wild boar, wolves and roe deer. He claimed that they were widely used, presumably, however, mainly in the Middle Ages and the 16th century. The poison was prepared from the roots of the Christmas flower (*Veratrum album*), which contain a strong toxin, *helleborin*.[81] It was best to pull up the roots at the end of August. They were washed, crushed and pressed in order to extract the juice. This juice was strained, boiled and painstakingly skimmed and then it was strained once more and set out in the sun from

---

80 Juan de Lastra is recorded as a crossbow maker working for King Philip IV of Spain (1621-65). A crossbow by him, bearing the inscription IN DOMINO CONFIDO, together with its accompanying bending lever survives in the Real Armeria, Madrid (Nos J96-7).
81 The usual English name for *Veratrum album* is White False Hellebore. It contains two closely related substances, veratrine and protoveratrine which are nerve poisons.

ten in the morning till mid-afternoon. This was repeated for three or four days until the sap was sufficiently concentrated. Apparently some people concentrated it by boiling, but this was thought to be less effective. When the sap had assumed the consistency of syrup the poison was ready. If one sniffed it, it induced violent sneezing (75:154-155). The steel heads of the poisoned bolts had long points of rectangular section with long, very slender necks between point and socket. The poison was painted on from the head to 5-6 fingers' breadth down the shaft. The part painted with the poison was wound with a ribbon of very fine linen. The poison acted like glue to hold the ribbon fast (75:154). It is possible that the use of poisoned bolts was introduced into Spain by the Moors. However, even before this, the Gauls had apparently been making use of poisoned arrows, and according to accounts from 1696 poisoned arrows were still in use in the French, Swiss and Savoyard Alps during the 1600s for hunting chamois (78:235).

The crossbows used in Holland during the first half of the 16th century were evidently of the same type as the weapons used in Norway (figure 55). It appears that even as late as the middle of the 16th century, crossbows with wooden bows were still in use in the Netherlands (113: figure 156).

Figure 47 Target crossbow with steel bow, palm rest, and English windlass attached. 17th century. Dutch or Belgian.

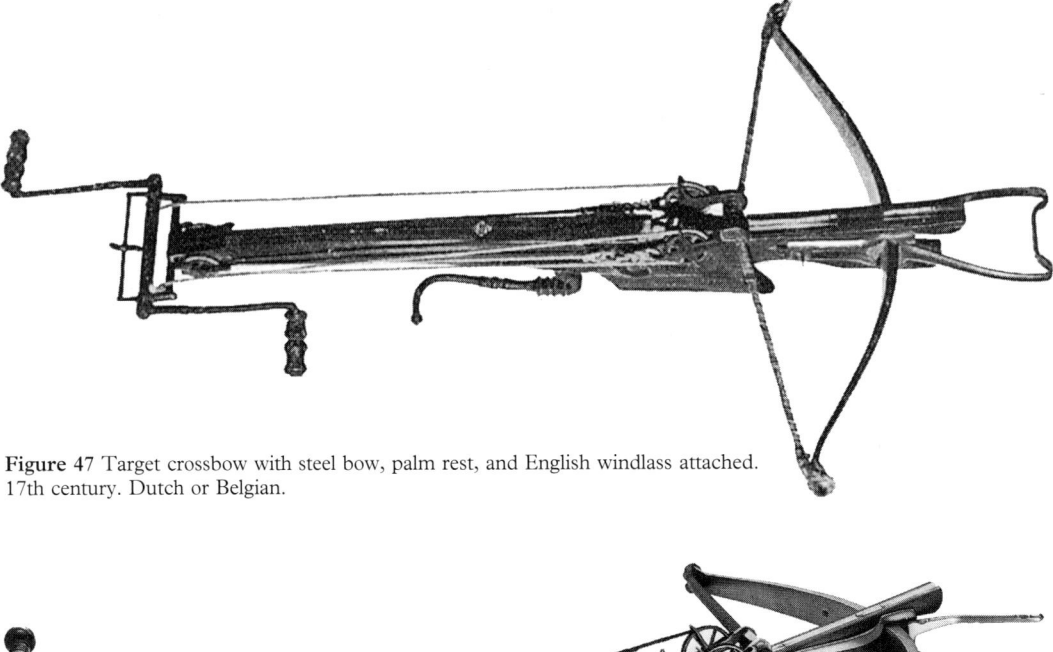

Figure 47a Target crossbow with steel bow, palm rest, and windlass attached. Western European 18th century. Royal Armouries, HM Tower of London, Inv. no.XI8.

Figure 48 Target crossbow with steel bow and support. Dated 1835. Belgian.

During the 16th century, small crossbows spanned by goat's-foot levers were also made in Holland (figure 46). Like the majority of 16th and 17th century western European crossbows they differ little in appearance from the types common in that area at the end of the Middle Ages. They have bolt-grooves, locks with nuts, and lever-triggers equipped with a spring and a safety catch in the form of a small pivoting arm on the underside of the tiller of the type that has already been described. They are reinforced with iron, but they differ from other western-European types in that the bow is held in place with bindings. A crossbow of this type preserved in the Livrustkammare has inlays of black and white horn and would appear to date from the middle or latter part of the 16th century. Another, in Consul General Jean Jansson's collection, is of exceptional quality with rich inlays of white and coloured horn and mother-of-pearl. Unlike the one described above, this crossbow had a bolt-retaining spring of iron, which was originally gilt, as were the side-plates, the trigger and the safety catch. The decoration dates the crossbow to the end of the 16th century, or just after 1600.

In Belgium and in some parts of Holland and northern France, crossbow shooting has been practised continually since the Middle Ages. Here there were many different types of crossbow, the majority never described, and among them old-fashioned forms and obsolete features held on for an astonishingly long time.

It is characteristic of the target crossbows used in these areas, during the period from the latter half of the 16th century to well into the 18th century, that they had a palm-rest immediately below the lock, as shown in figure 47 (65:168). In shooting, one grasped the rest with the left hand. Indeed, there was already a suggestion of such a rest in the western European crossbows of the Middle Ages (figure 20).

The crossbow illustrated in figure 47 dates from the 17th century, but is largely of medieval type. However, the lever trigger has been developed into a long trigger guard, and the lock has a set-trigger and is, in principle, designed in the same way as that illustrated in figure 38. Another Franco-Belgian type, is much like that in figure 47 just described, but instead of having a palm-rest, almost the entire underside of the tiller is swept out in a bow. Crossbows of this type, too, have hair triggers and are spanned with the English windlass.

It was in the 18th century in Holland and Belgium that the palm-rest to which we have referred grew into a support, often richly ornamented and carved (see figure 48,) the lower end of which afforded a good grip for the left hand. There are a number of different varieties of crossbows equipped with supports, presumably all typical of different dates and localities. The undoubtedly Belgian crossbow in figure 48 is dated 1835, but it exhibits some very old-fashioned details. The steel bow is 69cm long, 5.25cm broad and 1.45cm thick at

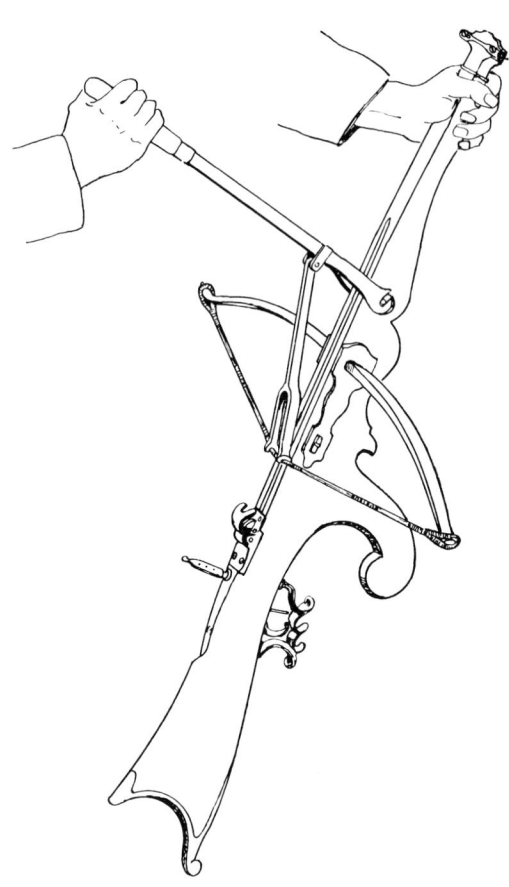

Figure 49 Modern Belgian target crossbow with spanner. After Payne-Gallwey.

the middle. The tiller has inlays of light-coloured wood shaped like butterflies and flowers. The bolt-groove is of staghorn. The bolt retaining spring, the nut and the nut recess are made of iron. The lock has a set-trigger of the type illustrated in figure 41. The crossbow is associated with a cranequin, which has an elongated bar cut on its upper surface with teeth, and a tiller loop of iron. According to Payne-Gallwey, cranequins with gearing positioned in this way and with this sort of tiller loop are of French type (75:144).[82]

In 1946 a number of crossbows with gaudily painted tillers passed through the Stockholm antiques trade. The tillers were mainly of the same type as those on the modern Belgium crossbow (figure 49). However, their butts, that is the part of the tiller behind the lock, were straight, and they reminded me very much of the corresponding part of the tiller on pellet crossbows of Italian type, such as that in figure 66. These crossbows presumably dated from the 18th century.

In Brussels, Ghent, Bruges, Antwerp and other Belgian cities, and in northern France and to a lesser extent in Holland, there are still crossbow-shooting clubs. Shooting at ordinary targets is the most common, although people also still shoot at popinjays or birds. In Brussels several firms still manufacture crossbows (75:206).

The modern Belgian target crossbow (figure 49), is heavy. The tiller is about a metre long. The bolt groove is only 30cm long, but from there forward the tiller inclines slightly downward so that the bolt does not touch it. The lock is a variant of the one illustrated in figure 41 and has a set-trigger. The trigger guard has rests for the fingers. The weapon has complicated sights, with a fore-sight right at the fore-end of the tiller. The bow is 68cm long, 2.5cm broad and 1.25cm thick at the middle. It is positioned 45cm behind the fore-end of the tiller and is held in place with iron bridles. Spanning is by a variant of the 'pusher' bending-lever illustrated in figure 44, this Belgian one (figure 49) being of iron with a wooden handle (75:208f).

The bolts used with this crossbow are 25cm long, weigh 40g and are fletched with three flights (75.218f). Payne-Gallwey says that with such a crossbow, and with bolts as described, he could put eight bolts out of 12 into the six inch centre of a standard archery target at 54m [60yds] range.[83] The other four bolts usually struck just outside the gold. Payne-Gallwey found that the extreme range was 225m [250yds] (75:206).

---

82 The affect of the teeth being on the top of the bar rather than on the side is that the wheel case and winding arm sit on the left side of the tiller, not on the top, and the winding is done in a different plane.
83 This is now not the size of the centre of a standard archery target, and there is some debate as to whether it ever was. W F Paterson thought that here Payne Gallwey was in error.

7 The Crossbow after 1500
The Scandinavian Countries

During the entire 16th century the crossbow was a very important hunting weapon in Sweden. It was also used in war until well into the 1570s, although its importance was gradually but constantly diminishing.

In the 16th century in Sweden, and probably also for a long time before that, the crossbow tiller was usually called *sula* (77:68, note 1), *armborstsula* (52:233), *sull* (2:225) or *sool* (12:149), all of which, as I have said earlier, derive from the Low German *zule* and High German *säule*.[84] The locks of these Swedish crossbows had a nut (12:149). In the Småland court records, the long lever-trigger was called a *trycke-nyckel* [literally 'press-key'] (52:233). In French, the trigger was called *clef* (79:657) and in German it was called *schlüssel* (118:45). Crossbow strings were called *skivegarn* (*skyffwogarn*) [literally 'slice-yarn'] (97:403) or *skivestreck* [literally 'slice-cord'] (67:140).

According to the well-known drawing by the German Landsknecht Paul Dolnstein (figure 50), in 1502 the peasants of the Västergötland had crossbows equipped with angular stirrups of western-European type (26:3). As I said in Chapter 4, this type of crossbow had been introduced into Norway and western Sweden through trade with the Netherlands.

It is very probable that primitive crossbows, like those from Skåne and Norway described in Chapter I, were used in the 16th century in many areas of Scandinavia as well as in southern and western Europe.

At the beginning of the 16th century crossbows with wooden bows are still mentioned in Swedish records. Olaus Magnus, who left his native country in 1524, says that pine (yew?) and larch were used for crossbow bows (71:10).[85] The Swedish edition of his book also speaks additionally of bows of *benved* [literally 'bone wood'] (70:36).[86] However, this is an incorrect translation, as the Latin edition mentions '*corneis ballistis*', horn crossbows (70:7).

It seems possible that at this time the majority of Swedish crossbows had horn bows. According to Dolnstein's drawing (figure 50), the West Göta peasants had such bows on their crossbows in 1502 (26:3). Olaus Magnus mentions horn bows as being in every way the equal of steel bows (70:85, 72:53).

On 10 December 1520 Christian II of Denmark issued an ordinance requiring all Swedish peasants to surrender their bows and crossbows (6:130).[87] Olaus Magnus reports that the owners were enjoined to burn their crossbows under threat of being charged with *lèse majesté*. Many thousands of crossbows were burned, although a number of shrewd people who had a presentiment of revolution hid their bows in the woods. Without crossbows, beasts of prey, particularly wolves, became an increasing danger to humans and domestic animals, and so the peasants had to find craftsmen and re-equip themselves at much increased prices, with many thousands of new crossbows to take the place of the ones burned. '*Yea, at times they even made bows glued together out of larch wood in place of horn*', Olaus

---

84 All these words mean 'pillar' or 'column'.
85 The Swedish words for pine and yew are *gran* and *idegran* respectively. The word used by Olaus Magnus is *grantrā*. Alm presumably felt that this should be translated as 'pine', but as a crossbow scholar would have felt that this was unlikely, and that Magnus may well have meant yew which was frequently used for the bows of crossbows as well as for longbows.
86 In England this tree is known as a spindle tree (*Euonymus europaeus*).
87 In 1520 Christian II defeated the Swedish regent Sten Sture, and took control of Sweden once more. This order was issued in an attempt to prevent further uprisings.

Figure 50 Prince Christian's landsknechts fighting against the peasants of Västergötland at Älvsborg in 1502. Drawing by the German Landsknecht Paul Dolnstein. [Note: Älvsborg Castle on the Göta Älv protected the harbour of Nya Lödöse (see Editor's Notes 61, 110, 112)].

Magnus recounts (72:52). Despite this statement the bows were more likely to have been made of yew or juniper wood rather than of larch, which almost certainly was not to be found in Sweden at this time.

Olaus Magnus says that the Goths and Swedes had powerful crossbows (72:52), and he asserts that *'the Goths surpass all other folk in shooting with their crossbows, seeing that never in the world have stronger crossbows been made than among them.'* (70:30).[88] It should be noted at this point, however, that Olaus Magnus himself came from Östergötland.

After crossbows with steel bows had come into more general use they were usually called simply 'steel bows' and crossbows with horn bows were called 'horn bows'. In the letters of Gustav Vasa, King of Sweden, one often comes across the expression 'steel bows and crossbows'. (42:264, 271, 302, 537, etc.). It seems safe to assume that the crossbows referred to in this way generally had bows of

---

88 By 'Goths and Swedes' Alm means the inhabitants of southern and central Sweden respectively (See note 128).
89 Jon Andersson was one of the leaders of a peasant uprising against Gustav I Vasa known as the Dacke feud (see note 92). 'Bråckenhussen' is a nickname of the Count Erik von Hoya who seems to be otherwise unrecorded. Alm may be referring to the German nobleman Count Johann von Hoya who came to Sweden in 1524 and in the following year married Gustav Vasa's sister Margareta. Amongst other tasks he was appointed by the King to negotiate with Lubeck about Sweden's debt with the City. The King refused to accept the outcome of these negotiations and Johann von Hoya left Sweden to join the King's enemies in 1529. Berend Von Mehlen, another German nobleman who entered the services of Gustav Vasa and married into the royal family, fell into disgrace after failing to capture Gotland in 1524 and fled to join the King's enemies, von Hoya, the Duke of Mecklenburg, the city of Lubeck and the King of Denmark.

horn. When Jon Andersson, leader of the peasants of Småland, was in Lübeck in 1538, he met Count Erik von Hoya, '*Bråckenhussen*', and Berend von Mehlen, and among other things he acquired from them seven bows of horn (20:22).[89] In 1544, Gustav Vasa directed the Governor of Stockholm to deliver a good, strong horn bow with all its accessories to a Finnish *vildskytt* [literally 'wilderness marksman'] who had found his way to Gripsholm Castle (43:652).[90] Despite this, it is obvious that at this time steel bows were becoming increasingly popular. One is recorded in the memorandum book of Jönköping for 1525 (57:110), and in Gustav Vasa's registry such weapons are mentioned for the first time in 1526 (37:225, 253). Olaus Magnus, too, refers to crossbows with steel bows (70:85). In an inventory dating from 1550 of the property belonging to Axel Eriksson Bielke's bailiff for Southern Vedbo and Ydre, a steel bow, a spanning-lever and a *köfver* (*koger*) [quiver] are included (85:115).[91]

In the so-called Växiö statute of 1530 Gustav Vasa directed that no one should carry a crossbow or other weapon to a town meeting, market, banquet or church, on penalty of 40 marks fine and loss of the weapon. However, he continued '*one may well carry such in cry after the wolf, the bear, and the fox, or in a squirrel grove*', but nowhere else (38:224).

The crossbow was used for a long time in Swedish armies. During the Dacke feud the King had directed that all soldiers should be armed with firearms, and this seems in fact to have been effectively enforced for troops in the field (7:128f).[92] However, it was staff weapons that the king had withdrawn and replaced with firearms, and the order is not thought to have applied to crossbows. In the 1540s, although the peasants of Småland appear to have made considerable use of firearms (7:129f), the crossbow remained their primary weapon (20:42, 48, 51).[93] The crossbows of Småland were of the usual Swedish type. Thus, for example, there is mention of a crossbow with a *krihake* in the Jönköping civic memoranda book of 1528 (57:120). As late as 1551 in Erik Mattson's Östergötland battalion 505 men had steel bows, whereas only 83 carried arquebuses and three were armed with other weapons.[94] In the same year, among Nils Eskilsson's Östergötland battalion 506 had steel bows, 51 had *rör* [arquebuses] 18 had *hakar* [hackbuts] a type of heavy hand-gun and 29 had either staff weapons or no weapons at all (76:77).[95] In the battalion from Dalarna in 1554 five men had hackbuts, 361 arquebuses and 301 steel bows, and 533 men-at-arms had horn bows (77:40, note 1). These horn bows were presumably of the same type as the weapon illustrated in figure 29.

Horn bows were clearly on the decline in the middle of the 16th century. As can be seen from the extract from the brigade rolls quoted above, there were no horn crossbows at all

---

90 Gripsholm Castle lies on the southern shore of Lake Mälar next to the town of Mariefred. A castle was first built on the site in 1380 by Bo Jonsson. The present castle was built for Gustav I Vasa by Henrik von Cöllen.
91 For Axel Eriksson Bielke see note 107. Sodra Vedbo is in Småland, Ydre in Östergötland.
92 The Dacke feud was a peasant uprising against Gustav Vasa which took place in Småland in 1542-3. The grievances which led to it were the high taxes being imposed, the protestant religion which was being introduced, and the ban on trade with Danish ports in Blekinge (the south-eastern most province of modern Sweden). It was led by the farmer, Nils Dacke, who had been outlawed in 1536 for murdering a royal official. He then joined the band led by Jon Andersson and fomented rebellion. The Dacke feud was probably the most dangerous of the uprisings against the King and, following a treaty signed in November 1542, Dacke was the virtual ruler of Småland during the winter of 1542-3. In the spring, however, the King sent a new army to Småland; Dacke was wounded and forced to flee across the border into Blekinge where he was later killed in August 1544.
93 Småland is a region in central and eastern southern Sweden, one of the main centres of the Swedish iron industry since the 15th century.
94 Östergötland is a region on the coast of southern Sweden just north of Småland. The Swedish word translated as 'battalion', *fanika*, (*Fahnlein*) does not have an accepted English contemporary equivalent. Literally it means 'small flag' and refers to a unit of about 500 men, which is sometimes part of a regiment.
95 'Hackbut' is the English term used at this date to describe a heavy military gun with a hook beneath the barrel, designed to be hooked over a parapet or stand to act as a recoil stop.

among the Östergötland men-at-arms. This almost certainly came about because the crossbow makers of Östergötland had already gone over to making crossbows with steel bows.

During the Russian war of 1555-7 firearms began to replace the crossbow in the Swedish armies for good. In addition, about one-third of the infantry were armed with staff weapons. Thus, for example, in the batallions from Dalarna all but three horn bows and also the majority of the steel bows, were supplanted by firearms and staff weapons. In 1556, Hans Svenske's Dalarna battalion was armed with 236 *gevär* [muskets], 56 steel bows, 3 horn crossbows, 32 halberds and 64 pikes. In Hans Pedersson's Dalarna battalion there were 183 muskets, 75 steel bows, 3 *fjäderspjut* [rawcons or corsèques], 25 halberds and 24 pikes (77:40, note 2).[96] In 1554 two steel bows were deposited in the armoury at Gripsholm (36:68), and in 1556 118 crossbows came into the Stockholm armoury (92:15). In 1560, in all Swedish armouries, there were 464 steel bows and 9 horn bows (92:4).

During the reign of Erik XIV, and part of that of John III, crossbows were still used to a limited extent in the Swedish army, primarily because of a shortage of firearms.[97] In 1563, there were in the various royal armouries 441 steel bows with hooks and levers, 12 horn crossbows and 5,221 hand firearms (2:233). In 1563, Länge Benkt's Västmanland battalion received 10 pistols, 122 guns of various sorts and 60 steel bows (50:51).[98] Also in that year, four of Mats Grot's contingent of soldiers from Uppland received steel bows with levers (17:17).[99] Similarly, in 1563 117 steel bows with *sula* (crossbow tillers), two steel bows without tillers and 17 levers (77:68, note 1)

were issued to a Dalarna levy, commanded by the bailiff Mikael Helsing, for use during an expedition to Norway. In 1564 the two Östergötland brigades received 86 steel bows (76:143). In 1571 427 steel bows were sent to Finland from Stockholm (76:209) and in 1573 223 steel bows arrived at Reval (76:216, note 37).[100]

In 1577 John III ordered the province of Nyslott in Finland to equip 1000 mounted bowmen (2:136). It was usually stated that such troops should carry crossbows. However, in a letter dated 12 November 1555 dealing with the equipping of cavalry in Nyslott province, Gustav Vasa ordered that the men should be armed with weapons that they could use on horseback, mentioning both guns and bows. It is probable that twenty-two years later bows rather than crossbows were again intended for the horsemen of Nyslott. In both cases it was evidently the mounted Russian bowmen that were the model.

In 1644 crossbows were carried by part of the Dalarna levy which besieged Särna and Idre under the leadership of the pastor Daniel Buscovius, the rest being equipped for the most part with muskets and bear-spears.[101]

At the beginning of the 16th century there were a great many *verkmästare* in Sweden, and they are quite frequently mentioned in civic Memorandum Books. Thus, for example, in 1506 Henrik, a *verkmästare* in Stockholm, was sentenced to pay a fine of 40 marks for manslaughter. As usual, the sentence was reduced and he escaped with payment of 10 marks in cash and four crossbows (l01:103, 107). In 1514, he received the more elegant title Henrik Balistarius (100:362) and in 1517 the council asked that he come to the Castle to

---

96 Rawcon, and corsèque are terms used to describe staff weapons with a three-pronged spear head. A rawcon more properly should refer to a type of bill.
97 Erik XIV reigned from 1560 until 1569 when he was deposed by his half brother, John, who reigned until 1592. Erik died in prison in 1577, traditionally said to have been killed by poisoned pea soup.
98 Västmanland is a region in central Sweden, west of Stockholm.
99 Uppland is a region on the east coast of Sweden, north of Stockholm.
100 Reval is a port in Estonia on the south side of the Gulf of Finland.
101 Särna and Idre are parishes in Dalarna on the border with Norway. In the 16th century they formed part of Norway.

settle a dispute over a crossbow between himself and Amund (101:148).

In 1519 there were five *verkmästare* in Stockholm (96). In the same city there is mention on 28 September 1517 of *'the commonalty of the* verkmästare *trade'* (the guild) in connection with a certificate stating that the *verkmästare* Per and his wife had drawn up reciprocal wills (101:164). The number of *verkmästare* seems to have fallen rapidly in Stockholm after this date and it is thought that in 1525 there were only two left (96). The last mention of *verkmästare* in the Memorandum Books of the city of Stockholm is on 23 November 1545, but the *verkmästare* Henrik who is mentioned there, and who may perhaps be identical with the one mentioned above, is recorded as having died in 1538.

At Tälje in 1500 there was a *verkmästare* called Jeppe (99:467).[102] In 1504 the council of Arboga ordered the *verkmästare* Håkan to pay three öre in coin to Måns the belt-maker and to deliver to him a new *sool* (crossbow tiller) with a new nut in it (12:149). At Jönköping there is mention in 1514 and 1517 of a *verkmästare* Anders (57:99, 106), and in 1515 and 1516 of a *verkmästare* Jens, who is also called *verkmästare* Jens Andersson (57:101, 103). The *verkmästare* Karl who is mentioned in 1526 and 1532 appears to have been the last to practise his craft at Jönköping (57:114, 124). The account books of the royal estates at Bergen record that in 1522 the castle's *verkmästare* received yarn for horn bows (18:80). The Hamar Chronicle, written between 1542 and 1553 records that at Hamar there were *'verkmästare who made horn crossbows'* (123:125), and that in 1537 Bishop Magnus had his *verkmästare* make a lot of large horn crossbows (123:143). There was a *verkmästare* at Malmö Castle in 1541 (18:83).[103]

In Copenhagen the last *verkmästare* seems to have been Mogens, who is mentioned in 1547 and 1548 (18:70), but as late as 1551 *verkmästare* are included among other craftsmen at Roskilde (18:86).

The *verkmästare* were succeeded by steel bowmakers, but these never became as numerous in Sweden as their predecessors had been. It is difficult to say when steel bows were first made in Sweden, but this probably happened around 1500. In 1518 the council at Stockholm directed Eskil the Smith to *'come to an agreement with the guild of smiths in the matter of the crossbow'* (101:202). In all likelihood Eskil had made a crossbow, probably with a steel bow, which was outside the normal purview of his trade. Eskil made many locks for the city of Stockholm, including those for the city gates. This is the one instance in which I have come across evidence about crossbows made by anyone other than a steel-bow maker or a *verkmästare*. The letters of Gustav Vasa show that in the 1540s there were steel-bow smiths *'at the frontier'* (43:658, 659), by which it is presumably meant southern Småland. Is it possible that the renowned gunsmithing industry of Göinge and Småland was preceded by the manufacture of crossbows?[104]

In the mining districts of Östergötland steel was produced in the 1520s and presumably earlier as well. Gustav Vasa recounts that it was manufactured in the parishes of Vångeberg, Tjällmo, Godegård and Hellestad (41:174, 175). It was presumably because of this steel making that the crossbow makers of Östergötland began to manufacture steel bows at such an early date. According to Gustav Vasa's letters there were a great many steel-bow smiths in Östergötland at that time. Two places are singled out for special mention as the

---

102 Tälje, now Sodertälje, lies just south of Stockholm between Lake Mälar and the Baltic. Its name changed when the town of Norrtälje was established, 150 kilometres to the north.
103 Malmö, now in the southern Swedish region of Skåne, was part of Denmark in the 16th century.
104 Göinge is in the south of the region of Småland near to the border with Skåne, and hence in the 16th century belonged to Denmark. The town and its region were renowned for their gun industry. Especially famous were the snaphance rifles of small bore produced there from the late 16th century into the 17th century which were known as 'Göinge-guns' or 'Småland-guns'.

homes of steel-bow smiths: Vadstena, where there nevertheless seem to have been only two such craftsmen (43:203), and the parish of Tjällmo.

The steel-bow makers of both Vadstena and Tjällmo were soon to feel the royal wrath. When Gustav Vasa was in Vadstena in the summer of 1543 he ordered a number of steel bows from the city's steel-bow makers. However, they were in no hurry to make delivery (42:575) and when the weapons finally reached Stockholm, the king found that the work was unsatisfactory. On 8 April 1544 he therefore wrote to the bailiff in Hov province, Nils Eriksson, stating that he had received at Stockholm a number of steel bows which the two steel-bow smiths at Vadstena had made for him

'and we have paid a great deal for them. So that we find it very certain that they have done us a great treason with these bows. For they have no range, and no sooner has one discharged a couple or shots with them than the bows are downright sprung.'

The bailiff was therefore

'to take these steel-bow smiths by the neck and send them well guarded to our governor in Stockholm and arrange that all their tools shall accompany them hither, whereafter we shall make trial of whether they cannot or shall not do better work for us up here than they have done down there,' (43:203-204).

An undated letter from the king to the governor of Stockholm, Botvid Larsson, which the editor of Gustav Vasa's correspondence dates without evidence, and apparently incorrectly, to 14 March 1544 (May is perhaps more likely) evidently deals with the same malefactors.[105] They had arrived at Stockholm, and the bailiff of the Castle had sent a letter to the king asking *'how he should deal with them'*. In his reply Gustav Vasa advised him that these steel-bow smiths

'made a number of steel bows for us like the rankest of traitors. And on that account it would be well merited if they were to be punished as becomes such rogues, yet one might try again to see if they are willing to show themselves of any use. Therefore you may let them be set to work, at the same time providing that adequate watch and supervision be instituted over them so that they play no knavish tricks' (43:131).

The steel-bow smiths at Tjällmo were evidently gentlemen who took no account of persons. They had made unserviceable crossbows and struck the Vadstena *verkmästare* stamp upon them. This gave rise to the following royal letter, dated 12 July 1543, which is interesting from the standpoint of arms history:

'In as much as we have understood that you, who practise the verkmästare *craft there in Tjällmo parish, also make steel-bows that are yet good for little and are false work, something which cannot by any means be condoned in good steel-bow makers when one comes to have knowledge of such things, etc. Thus seeing we have become cognisant of such falsity in your work, both in that you strike the marks of these Vadstena* verkmästare *upon your steel bows, and in that the work is in itself unacceptable and intolerable, we shall on that account by no means suffer you after this day to make any more steel bows. And we order you as you value your necks to conform to this, unless you are willing to move here to Vadstena and become citizens here among these other* verkmästare *so that it may be made plain what work you do and how well, thus not deceiving folk with your steel bows as you have done hitherto. Be governed by this'* (42:411).

From this royal letter it is apparent that *verkmästare*, too, could manufacture steel bows.

---

105 The editor of the correspondence of Gustav Vasa referred to was working in the second half of the 19th century (see Alm references 35-46).

Gustav Vasa appears to consider that the crossbow makers at Tjällmo would be able to do better work if they moved to Vadstena and came under the control of the *verkmästare* or the steel-bow smiths living in that city. It was presumably these latter who had brought the matter to the King's attention. If that was the case, nemesis was on watch, for the following year the steel-bow smiths of Vadstena came in for very rough treatment, as we have just remarked.

However, the steel-bow smith situation was critical, and the king had great need of such men. Perhaps not all of the steel-bow smiths of Tjällmo had sinned, or perhaps the accusations had been exaggerated. On 27 January 1545, the king wrote to his mother-in-law Ebba Eriksdotter Vasa, widow of Erik Abrahamsson Leijonhufvud:

'We have several times in the past written for steel-bow smiths there in Östergötland, that they should betake themselves up here to us, something which has however not yet taken place, wherefore it is our will and dear desire that you with your bailiffs there in Tjällmo bring it about that the steel-bow smiths who are resident in Tjällmo betake themselves with their tools up here to Stockholm without delay; and the bearer of this letter, Peder Larentzson, will be able to advise you further which men these are' (44:62f).

Presumably the parish of Tjällmo was part of Lady Ebba's fief.

On 1 June 1545 Gustav Vasa sent a letter to Måns Nilsson, governor in Kumogård province, stating that he had written to Östergötland for some good steel-bow smiths and that he was thinking of sending some of these to Finland so that they might there make *'a number of good steel bows, in order that some good weapons may reach even that extremity of the nation'* (44:343-344).[106]

In the king's letters to Axel Eriksson Bielke there is mention of deliveries of steel bows to the crown in 1546 and 1547.[107] At that time, Lord Axel had as his fief the greater part of Bankekind district plus the whole of the Ydre and Southern Vedbo districts.[108] On 28 July 1546, the king wrote that he had received a letter from Bielke in which the latter apologises for *'some of the steel bows not being all that strong'*. The king stated that he could not make out how the matter stood or what Hans Tomasson (governor in Småland) had to do with their making. The king also insisted that if the smiths *'did indeed receive something more than one half pound* (a half, or a pound and a half) *of steel for the making of each bow'*, the weight be recovered when the bows were delivered rather than be written off.[109] Gustav Vasa closes with a rather sharp admonition that the steel bows should be made as they ought to be made and as the king had earlier communicated to Bielke both orally and in writing (45:114). To judge by the above, these steel bows would seem to have been manufactured in Småland. Lord Axel had evidently sought to skimp on the steel.

In 1547 Axel Eriksson Bielke informed the king that he had sent a number of steel bows to Strängnäs. In his answer, the king raised doubts as to whether the bows were *'so well finished and good as rightly ought to be the case'* (46:500).

There were steel-bow smiths in Hälsingland as well (122:7). Gustav Vasa probably got steel-bow smiths brought up from Östergötland

---

106 Kumogård was one of the Swedish provinces of Finland.
107 There appears to be no evidence that an Axel Eriksson Bielke existed at this time. Alm almost certainly refers to Erik Axelsson Bielke, in error transposing the first name and patronym. Erik Axelsson Bielke (c.1500-1559) commanded the royal army in the Dacke Feud, and, by signing a treaty with Dacke which the King disliked, lost many of his estates in 1544. He was married to Anna Eriksdotter, sister of Queen Margareta Leijonhuvud, and daughter of Lady Ebba Eriksdotter Vasa.
108 This area equates to the southern part of the province of Östergötland and the northern part of the province of Småland.
109 Alm was unsure how to interpret the King's letter. It might be interpreted as reading 'half a pound' or 'a pound and a half' or steel. However, the latter seems more likely to be the correct weight.

to Stockholm. Among these was probably one steel-bow smith called Lasse, who is mentioned in the Revenue Office Book for 1546 and who, in 1547, delivered 35 bows to the royal armoury. Subsequently he became a burgher of Stockholm, taking the burgher's oath on 2 April 1554 (102:56). Amongst others he worked for Duke Erik, and as late as 1570 he was executing orders for the royal family (73:7). On 15 May 1578 he resigned as a burgher (103:365), and on 8 August 1580 it was reported that he was unable any longer to practise his craft because of his advanced age (104:155). Lasse probably made crossbows based on fashionable Nuremburg and Augsburg models. He is the only steel-bow smith to be mentioned in the Memorandum Books of the city of Stockholm for the whole 16th century.

During Gustav Vasa's reign several foreign steel-bow smiths emigrated to Sweden or were invited there. On 31 December 1543, Gustav Vasa wrote to Erik Fleming that he had learned that a good steel-bow smith had come to Sweden, and he directed Lord Erik to find him *'a good and convenient position at some one of our castles'*, so that he might make a number of steel bows for the king (42:596). On 12 December 1545 the king wrote to Gustav Olofsson Stenbock, governor at Älvsborg, and directed him to organise the import into Sweden of some good steel-bow smiths *'from the westward'*.[110] The king had in fact got himself some good steel smiths who were making for him *'the noblest sort of steel such as men could make no better any place in Germany'*. These steel smiths were employed in the iron works in Östergötland (44:583). Western Europe is suggested by 'from the westward', and probably most came from the Netherlands. In a letter dated 27 February 1546 to Peder Skrivare, bailiff at Linköping, the king stated that the peasants in the Hanekind district had promised to deliver 12 cartloads of charcoal for the steel-bow smith's use (45:21).[111] It is not known whether this steel-bow smith was a Swede or foreigner.

Crossbows were also imported. In a letter of 14 October 1543, the king directed Axel Eriksson and Gudmund Skrivare to deal firmly with *'country merchants'* in the Kalmar area and elsewhere, who were carrying on unlawful trade and who did nothing else than spirit provisions and other forbidden articles out of the country and in return merely gave the peasants steel bows, crossbows and other weapons, *'of which pernicious weapons the peasants otherwise have little, or no use'* (42:537). It was presumably this sort of trade that Olaus Magnus had in mind when he said that craftsmen in the *'Low German areas'* manufactured powerful crossbows (*ballistae nervosae*), which were exported to the northern countries and either sold there or exchanged for oxen or horses (70:2). By 'the Low German areas' he probably meant both northern Germany and the Netherlands.

Gustav Vasa had himself imported crossbows with horn bows from Lübeck (67:138), and on 5 March 1544, he ordered Gustav Olofsson Stenbock, governor at Älvsborg, to procure 100 or 200 strong steel bows *'such as they make to the westward'* (43:111). Gustav evidently asked what the specifications of the crossbows should be, for on 14 March the king answered that some of them should be made so strong that they could be spanned with *krihake* and back, and some so strong that they could be spanned with strong pump-handles *'and you should have the bows made rather long and otherwise in every respect well finished'* (43:134). Stenbock seems to have been in no hurry with the order, for on 14 October 1544 Gustav Vasa sent him an impatient reminder in which he remarked, among other matters, that he had written twice about the ordering of the good, strong, western bows and that Lord Gustav

---

110 The castle of Älvsborg now lies under the city of Gothenburg near the southern end of the Älvsborg bridge. It was built in the 14th century to protect trade, and eventually was demolished in 1660 and replaced by a new fort, Nya Älvsborg, on an island in the estuary of the Gota Älv.

111 Linköping is one of the major towns of Östergötland; in southern Sweden. It lies some 20 miles east of Lake Vättern. The Hanekind district lies just south of Linköping.

Figure 51 Interior of a crossbow maker's workshop. After Olaus Magnus.

must not let the matter slip his mind (43:661). According to the register of imports, among items imported at Nya Lödöse in 1546 were eight dozen steel bows, three lever-spanned bows and a footbow, this latter presumably also being some sort of crossbow (124:276).[112]

In 1546, the price of an imported steel bow was set at five marks, and that of a horn bow at four marks (73:266). According to an ordinance of the 1570s, a steel bow imported from abroad, with its bender or hook, *'should cost 17½ daler in Sweden'* (73:267). Arms and armour were also free of duty in the reign of Gustav Vasa (73:8, note 4).

Swedish crossbows in the 16th century were usually spanned with a *krihake* or a spanning-hook. Olaus Magnus illustrates the interior of a crossbow-maker's workshop, with a spanning belt with spanning-hook, and also another spanning-belt with a spanning-strap and double *krihake* hanging upon the wall (figure 51) (70:29). According to the work tariff established for the city of Stockholm by Gustav Vasa in 1546, a *'spanning-belt'* should fetch three öre and one örtug [tenth], a *'spanish belt'* should cost the same amount, but a *'Deyen-bälte'* [sword-belt] should only cost one öre (61:318). A spanning-hook is mentioned in 1504 at Stockholm, for example (100:8), and in 1506 at Arboga (12:181). In a market price tariff assessment for trade with the Lapps in the Umeå and Ångermanland reserves dating to the middle of the 16th century or a little later a *'good crossbow with hook'* is valued at two marten skins (120:28).[113]

The *krihake* was in very general use. Nordiska museet [The Nordic Museum] has a double one dated 1548 (25:81). Olaus Magnus wrote: *'No less highly valued are strong crossbows with their associated pulleys by means of which they are spanned with astonishing speed by bending the body over them'* (70:30). As was remarked above, goat's-foot levers and pump-handles, were used in Sweden in the mid and late 16th century (43:134, 85:115, 77:68, note 1). Crossbows spanned with a cranequin were not uncommon in Sweden during the 16th century, but they are seldom mentioned. However,

---

112 In 1473 Nya or New Lödöse was built to replace Lödöse (see note 60) on the site of what is now Gamlestaden (Old Town), a part of modern Gothenburg. Later the inhabitants of Nya Lödöse were moved from the area of Gamlestaden to near the Old Älvsborg fortress (see note 110) for their protection. This did not prevent the town being destroyed in 1619 by the Danes.
113 Umeå is the chief town of the region of Västerbotten and lies about three miles from the Gulf of Bothnia on the eastern coast of Sweden, some 300 miles north of Stockholm. The region of Ångermanland lies immediately to the south of Västerbotten.

in 1521 it is recorded that a Swedish cavalryman, Little Bengt, who was a man-at-arms (and thus a professional soldier), carried a cranequin (107:39). *'Hooking up'* meant spanning a crossbow, and this expression was used as a word of command in Sweden during the 16th century (20:34).

As has been mentioned earlier, a crossbow with its spanning mechanism and quiver, presumably filled with bolts, was called a *'crossbow with all its shooting kit'* (100:70). In Sweden in the 16th century military bolts were of the same types as were used in the preceding century, that is to say some were of the international type with socketed heads, and some were Dale bolts. During the unsettled years 1501 and 1502 at least 12 casks of iron were forged into bolt heads for the city of Stockholm (95:32,33,36, etc). On 28 February 1502 a single payment was made to smiths for no less than five casks of iron that had been forged into bolts, and it was remarked that *'400 dozen bolts came of it'* (95:31). In 1518, too, a number of casks of iron were made into bolt heads for the city of Stockholm (96:74-77). Most frequently it was ordinary blacksmiths who forged the iron into bolts, but in 1502 (95:33) and 1505 (95:125) Peter the Cutler is recorded as having carried out such work, and on 6 June 1502 one Nils the Locksmith received smith's pay for a cask of iron that he had forged into bolts (95:36).

In Stockholm in the days of Sten Sture the Younger there were two professional boltmakers, namely the Anders *Pilstickare* referred to earlier and also Nils Persson *Pilstickare*. In 1518 the city of Stockholm bought great quantities of bolts. On 23 August the bolt-makers were paid for 116 dozen bolts (96:77). Nils Persson *Pilstickare* was the biggest supplier with at least 253 dozen (96:75, 79, etc). In the first half of 1520, Nils Persson received at least 45 marks for bolts in a number of instalments (96:123,124,156,157), and on 11 August he got 8 marks 3 öre for 144 dozen bolts (96:161). Nils Persson *Pilstickare* and Anders *Pilstickare* are the only craftsmen of their type to be mentioned in the Memorandum Books of the city of Stockholm. The latter was still alive in 1525 (96:266). It is not usually indicated whether the bolts bought by the city of Stockholm were of the international type, equipped with socketed heads, or of the older model, equipped with tangs and known as Dale arrows. However, the accounts at the beginning of the 16th century are an exception, as they do refer explicitly to Dale bolts.

Although heads of Dale bolts were forged in Stockholm (95:33), in 1504 sixty-five dozen unshafted Dale bolts were purchased (95:80). In 1508, the burghermaster Jöns Gudmunsson procured sixty-one dozen Dale bolts (95:274), and in 1510 seventy-two dozen Dale bolts were bought from Olof Eriksson at Tomta village in Romfartuna parish (95:356). During the conflict between Sten Sture the Younger and Archbishop Gustav Trolle, the Dean of Forsa in Hälsingland had 20,000 Dale bolts forged for the Archbishop (49:972).[114] Heads of Dale bolts have been found in the ruins of the archiepiscopal fortress at Almare Stäk, which was demolished in 1517.[115]

Dale bolts were not only used in Sweden. When Bishop Magnus of Hamar took up arms in 1537 against Christian III's captain Truid Ulfstand, *'he sent bids to his friends and merchants in Sweden for 6,000 Dale bolts, which he did indeed secure'* (123:143).[116]

In 1521, Gustav Vasa taught the people of Dalarna a new way of forging bolts *'with not so long a point and with spirally-set flights, so that*

---

[114] Sten Sture the Younger was Regent of Sweden from 1512 to 1520. Archbishop Trolle supported the Danish Ring, hence their conflict.
[115] Almare Stäk was Archbishop Trolle's Castle on Lake Mälar guarding the waterways to the cathedral city of Uppsala. Built in the 14th century it was destroyed in 1434, then rebuilt, and finally demolished in 1518 following Archbishop Trolle's surrender to Sten Sture in 1517.
[116] Hamar is a town in Norway on the shore of Lake Mjösa north east of Oslo. Christian III reigned over the united kingdom of Denmark and Norway from 1533 to 1559.

Figure 52 Capercaillie hunting with crossbow and 'stalking horse'. After Olaus Magnus.

they could hook themselves on and screw their way through armour' (107:21). He was quite simply teaching them to make bolts of the international type described above with a short, socketed head and with curved fletching (*'sneda slindor'*) (figure 17). Bolts of this type normally weighed 60-75g each, and therefore at the range usual in crossbow shooting they were of greater effect than the Dale bolts, which normally weighed from 35 to 45g.

Per Brahe the Elder makes an interesting comment on the bolts of the Småland men. He says that after a fight with Dacke's troops he had his arms and legs entirely covered with bruises from the bolts '*on which the* hullangarna [barbs] *were loose*' (20:48).[117] In a variant of his chronicle the word is *hållungarna* (20:71). Presumably Per Brahe is talking about the long heads of Dale bolts. During the campaign he wore armour, and he had a black jacket over his harness (20:42). The material over his arms and legs must certainly have had a number of dents from hits by bolts.

The *skäktor*, or broad-headed bolts, are rarely mentioned. In the Memorandum Books of the city of Stockholm this class of bolt is mentioned only once, in 1511 (100:211). In an arms ordinance of 1531 every soldier that silver miners and copper miners were asked to provide to meet an attack by Christian II was ordered to bring with him a good crossbow and four to six dozen bolts or *skäktor* (39:455).[118] Christian II's daughters had demanded dowries from Sweden, but the parliament that met in 1547 absolutely refused to consider the idea and declared that if it were to become necessary they would have forged '*for those queens subsidies and gifts of good bolts*, skäktor *and spadar, and that they would add a gratuity of gunpowder and lead*' (20:57) As has been suggested above, a *spada* was a broad-headed type of bolt-head, yet one very seldom mentioned.

The tactics of the Swedish peasants to shoot bolts obliquely upward against cavalry (70:43) and vehicles (70:194, 260), (figure 32), so that they fell downward upon the enemy like a hailstorm, was obviously used at Brunnbäck Ferry, if one may believe the well-known lines:

The men of Dalarna fought forward to Brunnbäck,
Where the Danes they had in their sight.
Soon in the air, there were more of their Dale-bolts
Than a hail that might fall from the sky.[119]

---

117 Alm's text here is obscure and the quotation impossible to interpret. It does not seem credible that it means that these bolt heads had pivoted or expanding heads.
118 Christian II reigned over the united kingdom of Denmark and Norway from 1513 to 1523 and over Sweden from 1520 to 1523. He was dethroned in 1523 and died in prison in 1559.

Figure 53 Popinjay shooting with guns. After Olaus Magnus.

During the 16th century the hunting bolts used were of the same type as those described in the previous chapter, namely: bolts with broad, razor-sharp heads for hunting bears and wolves; bolts with sharp points for birds; and bolts with blunted heads for squirrels and martens. At Stockholm a bolt of the last type, a *boltekolv* ['piston-bolt'], is mentioned in 1511 (100:211). Later such a bolt was usually called simply a *kolv* ['piston']. From Småland there is mention in 1609 of a *'tipped piston'*, evidently a sharp-pointed bolt, that a man had shot into the arm of an enemy sixty years before (52:233). In 1594, Måns Olofsson Stiernbielke tells of a serving-man in Ydre who *'feathered pistons'* and in doing so made use of a glue-pot, and caused a fire that destroyed a farmhouse (85:116). Olaus Magnus has a picture of capercaillie hunting with crossbows, where the hunter creeps forward behind a specially trained *'stalking horse'*, figure 52. To judge by this picture, the hunting dog was trained to keep an eye on the bolts discharged and to bring back those that missed (72:121).

Per Brahe relates that one of Dacke's lieutenants was discovered by the king's men *'as he stood with spanned crossbow and shot at a hare, which he hit'*, but just at that moment he was shot himself (20:51).

According to the woodcuts in Olaus Magnus, when shooting one usually laid the butt of the crossbow tiller against the shoulder or the cheek (71:128, 133; 72:121). However, in the picture that shows the shooting of Dale bolts against cavalry (figure 32), the marksmen are holding their crossbows in front of them with both hands, the end of the butt being at thigh level, and shooting obliquely upward. We may assume that this shooting position is correctly represented as, for shooting obliquely upwards, similar positions are also mentioned at a much later date (1:221).

Target shooting and shooting competitions appear to have been very common in the 16th century. Olaus Magnus says that the Goths had always practised shooting and that no other people had greater experience in shooting with the bow (71:131). He also has a picture of target shooting at a disc in which blunt-headed bolts are used as usual and trained hounds bring the bolts back (71:128); and another picture of target shooting with guns, and popinjay shooting with crossbows equipped with blunt-headed bolts (figure 53) (71:133).

Olaus Magnus gives the following account of shooting competitions:

---

119 The battle of Brunnbäck Ferry took place in April 1521 when the men of Dalarna beat a 'Danish' army, commanded by Jens Beldenek, made up of troops from Denmark, Holstein, Scotland and France.

'Among the dwellers in the north, just as among innumerable other folk living near to them, it is a generally accepted custom that during the early summer all marksmen, both from the towns and the countryside, gather on free days or festival days on ground that has been specially prepared for this purpose, in order to demonstrate publicly their skill in all sorts of shooting. For this they may use either guns or crossbows or bows according to their own choice, and with their bolts or bullets they must hit certain targets set up in the open field. However, for the most powerful crossbows the target consists of an artificial popinjay, which sits at the top of a rotating iron pole. This they seek to dislodge with blunt-headed bolts, so that it falls down. After the end of the competition those in charge are wont graciously to give the most skilful a good prize, and at the same time to grant him some special right of priority, so that in future he can take his place among the highest ranking burghers. It is customary for the others, too, in conformity with the degree of their skill, to be encouraged with appropriate liberality and to be considered for future meetings' (71:133).

In the 16th century the aristocratic and elegant sport of popinjay shooting had spread even as far as the Swedish peasantry. In 1542 the governor at Västerås confiscated from the Färnebo guild a popinjay with chain, all made of silver and weighing 8½ lod [half-ounces] (3:44).[120] The king had embargoed the chattels of all the guilds, and thanks to the revenue-raising zeal of his governors there are today no popinjays of this period surviving in Sweden.

Figure 54 Quiver of elk-hide for crossbow bolts.

Crossbows were used, primarily for target shooting but perhaps also for hunting, by the nobility and other distinguished people in Sweden until well into the 17th century. The crossbows that the steel-bow smith Lasse, who was mentioned earlier, delivered to Erik XIV and Johan III could well bear comparison with the best foreign ones.[121] In 1608 Jakob Lådemakare (gun stocker) at Norrmalm received 12 öre for gluing the tiller of a steel bow intended for Duke Gustav Adolf (73:284).[122] Queen Christina owned at least two small crossbows, dated 1646, which can scarcely have been used for anything other than target shooting.[123] Presumably, they were manufactured for her use, as amongst the decoration on the tiller is a scene of a lady shooting at a target with a crossbow (Livrustkammaren 2434-2435).

120 Västerås is the capital of the province of Västermanland which lies due west of Stockholm at the western end of Lake Mälar. The governor referred to was the provincial governor. Färnebo is a parish in the province. The silver popinjay refered to here would not have been a target, but probably either the badge of office of the Master of the Guild, or a shooting prize.
121 Erik XIV (born 1533), reigned over Sweden from 1560 until 1568 when, following his marriage to his seventeen year old concubine, he was deposed by his two half brothers, one of whom succeeded him as Johan III. Erik died, it is believed by poison, in 1577. Johan (born 1537) reigned until his death in 1592.
122 Norrmalm, then a northern suburb of Stockholm, is today in the centre of the city. Duke, later King Gustav Adolf, usually known in English speaking lands as Gustavus Adolphus, reigned from 1611 until his death at the battle of Lützen in 1632. He was a brilliant military commander, and his intervention in the Thirty Years War in 1630 and his subsequent military successes raised Sweden to the ranks of the major European powers.
123 Queen Christina, daughter of Gustavus Adolphus, was only five years old when he died in 1632. For twelve years Sweden was ruled by a regency then, in 1644, on her 18th birthday, Christina began her majority rule. She was crowned in 1650, and abdicated in 1654 in favour of her cousin, who ruled as Charles X Gustav.

In Denmark in 1555 Christian III issued a document written in German, appointing Elias Klüchtzersz as the royal crossbow maker (*Armbrustirer*).[124] Hans Schenk was similarly appointed by Frederick II in 1571.[125] Such an *Armbrustirer* had the duty of storing and repairing the king's crossbows. In 1602 Kristoffer of Augsburg, a firearms marksman at Kronborg, was named *Armbrustirer* with the duty of looking after Christian IV's crossbow cabinet (18:82).[126]

In the middle of the 16th century crossbows began to be supplanted by guns as hunting weapons in Sweden as well. As early as 1532 Gustav Vasa says that in Värend there were men who could handle firearms (40:46), and by the middle of the century guns were being manufactured on a large scale for the king by country smiths in Småland, Närke, Uppland, Hälsingland, etc.[127] Undoubtedly these smiths also made hunting weapons.

Presumably in Småland, but certainly in northern Svealand, Norrland, Lapland, Finland and parts of Norway, crossbows continued in use for a long time, in some cases into the 19th century.[128] This was because they were particularly suitable for hunting squirrels, which was common in these areas. Naturally, only bolts with blunted heads, '*pistons*', were used for such hunting. However, it is probable that sharp-pointed bolts were also used to a certain extent for other types of hunting. Powder and lead were expensive then, and were often difficult to obtain. A few quivers of elk-hide have been preserved (figure 54) and these suggest the use of sharp-pointed bolts, as '*pistons*' were usually carried stuck into the belt.

It is possible that during the second half of the 16th century and the 17th century the peasants of Småland carried crossbows that were spanned with the cranequin. This is suggested by the comparatively frequent appearance in Småland both of this sort of spanning apparatus and of trap crossbows that have been altered from hunting crossbows spanned with such a device.

The typical crossbows of northern Svealand, Norrland, Lapland and Finland usually have steel bows 75-90cm long which are 4-5.5cm broad and 1-1.75cm thick at the middle. The majority of the tillers are 75-95cm long. The distance between the unspanned string and the rear side of the spanning gap is usually 15-16cm and seldom longer. The lock consists merely of a nut and a long lever trigger. The weight ranges from 5-8kg. Crossbows of western European type are usually a little heavier than the others.

When in the middle of the 16th century people in Dalarna began to replace their horn crossbows by crossbows with steel bows, the majority of the new weapons seem to have been imported from Norway, and because of this the western-European type of crossbow became

---

124 Christian III, King of Denmark and Norway 1533-39.
125 Frederick II, King of Denmark and Norway 1559-88.
126 Kronborg Castle at Helsingör, Denmark, is best known as the castle of Shakespeare's Hamlet. It lies at the north east of the island of Zealand commanding the northern end of the Oresund which separates Denmark from Sweden. Christian IV was King of Denmark and Norway from 1588 to 1648.
127 Värend is a part of southern Småland which lies south of Lake Vättern. Närke is a region in central Sweden south west of Stockholm between Lakes Mälar and Vättern. Uppland is a region north of Stockholm on the east coast. Hälsingland is a region lying further north on the east coast of Sweden.
128 Svealand and Norrland are two of the three historic parts of Sweden, the other being Götaland, each traditionally separated from the others by large tracts of forest. Götaland, the land of the Goths, lay around the Lakes Vänern and Vättern. Svealand, the land of the Swedes consisted of the area around Lake Mälar and Stockholm. Norrland, the northern lands, composed the northern coastlands on either side of the Gulf of Bothnia, thus including areas now in Finland as well as those in Sweden. Norrland emerged as a distinct district in 1433 and later areas to the north, west and south were attached to it, including Lapland. However some continue to believe that these new areas do not really belong to Norrland, hence Alm's mention of both Norrland and Lapland. Another reason for the survival of the crossbow in Norrland is that in 1664 the right to hunt in southern Sweden (i.e. Svealand and Götaland) was restricted to members of the nobility, whereas it was left free for all in Norrland (where very few noblemen lived). This also helps to explain the survival there into the 19th century of traditional snaphance-style rifles, some of the later ones even being made to use percussion caps.

medieval Swedish crossbow, which was closely related to the northern-German type (figure 29). The tillers have rounded contours. The fittings and the inlays are of reindeer or elk horn. The upper surface of the tiller between the lock and the fore-end is covered with a broad and thick strip of horn which continues beyond the bow. For the greater part of its length this strip is flat and it does not have a bolt-groove at the back. At the front the strip is a little swept upward, and here there is a short bolt-groove. On the underside of the tiller there is also a broad and thick strip of horn and this too continues beyond the bow. The horn strips applied to either side of the lock are of considerable length. They are joined together by a number of iron rivets as are the strips on the top and bottom of the tiller. The front of the nut recess is lined with horn. On the top of the tiller behind the lock there is almost always a horn strip of elongated triangular outline, which in some examples extends all the way to the butt. The fittings and inlays are normally decorated with a type of chip-carving. The tillers are far

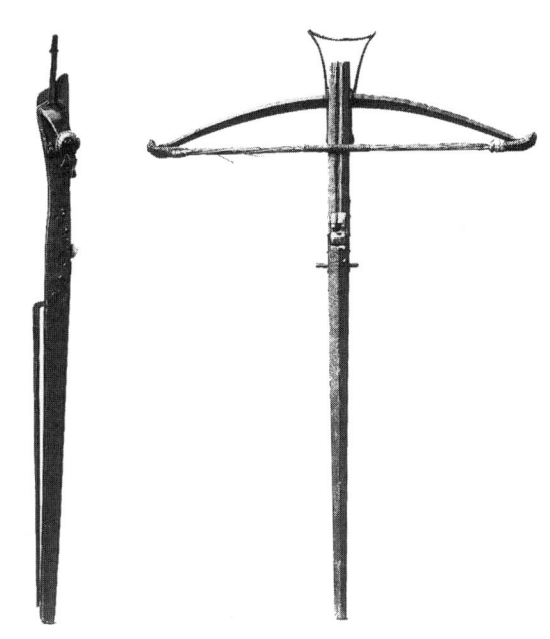

Figure 55-a Crossbow with steel bow, spanned with goat's-foot lever. Dutch type. Norway and Dalarna 17th and 18th centuries. Livrustkammaren [Royal Armoury], Stockholm.

predominant at that time in this area. The crossbows of western-European type, which were common then in Dalarna, Jämtland and Härjedal (figure 55), often have stocks with angular outlines.[129] The stirrups are also angular. Most of them have bolt-grooves, which in many cases are cut into an iron or horn strip, which is inlet into the upper side of the tiller and which projects a little beyond the forward end of the tiller. All fittings, including those that retain the bow and the stirrup, are of iron. Many of these crossbows are made to be spanned with the goat's foot lever, and thus usually have a small stirrup (figure 55). Dale crossbows that were spanned with the *krihake* frequently have hexagonal stirrups (figure 56). In addition, there are a number of transitional forms of the Swedish crossbow types described below, but these nearly always have fittings of iron not horn.

The crossbows used in Norrland, Lapland and Finland (figure 57), derive from the late-

Figure 55-b Crossbow with steel bow, spanned with goat's-foot lever. Nut missing. Scandinavian imitation of Dutch type. Norway and Dalarna. 17th and 18th centuries. Livrustkammaren [Royal Armoury], Stockholm.

---

[129] The region of Jämtland lies just south of Lapland, bordering Norway. The region of Härjedal lies immediately to the south of Jämtland.

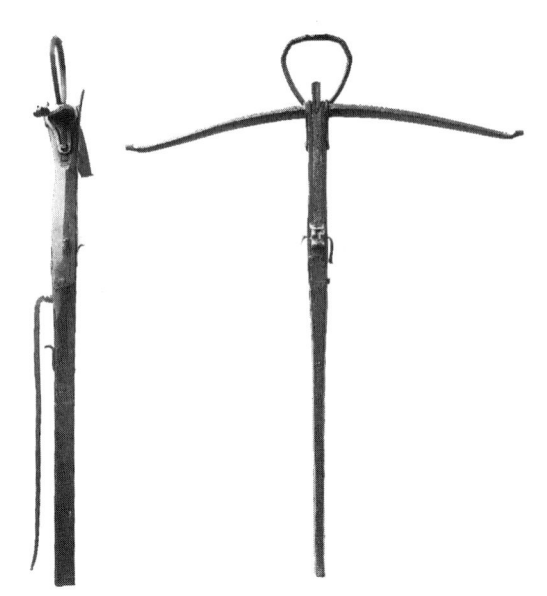

Figure 56 Crossbow with steel bow, spanned by a Samson belt. Transitional between Dutch and Scandinavian types. Probably from Dalarna. Livrustkammaren [Royal Armoury], Stockholm.

more sturdily constructed than those of medieval crossbows. The bows are held fast to the tillers by means of lashings of cord, sinews or narrow leather straps. The stirrups, which are for the most part of rounded form, are lashed fast to the bows. The cord is always of extremely fine hemp thread, and now and then it is twisted. As is usual, it is reinforced at the middle and at the loops with a binding of strong thread. On late crossbows, the string is sometimes slack. The reason for this seems to be that whoever made the string did not have access to suitable apparatus for putting it on in the right way.

Most of the steel bows for the crossbows used in Norrland, Lapland and Finland during the 17th and 18th centuries were certainly made by skilful rural smiths. Castrén related that, in 1754, wrought iron and an exceptionally good steel were made by ordinary smiths in the Sotkamo and Paldamo parishes in northern Finland (23:66, 67, 68, 71). It does not appear that the Swedish peasant smiths in any way

lagged behind those of Finland. If one had access to good steel it was presumably no more difficult to forge a bow than, for example, a scythe. However, to harden and temper the bow in the right way is said to have been very difficult.[130] Nevertheless in the 17th and 18th centuries steel bows are reported to have come to the areas mentioned above from central Sweden, Hälsingland and Norway (122:7).

Tillers, locks and stirrups were manufactured by the peasants. It was presumably not easy for any one to make a useable tiller and its lock, and even among the Lapps, certainly amongst the majority of the forest and fisher Lapps, crossbow tillers were made by particular individuals. In the 18th century the mountain Lapps rarely had guns or steel bows, but naturally enough the forest Lapps and the fisher Lapps were usually very good shots, both with the crossbow and with the gun.

In Norrland, Lapland and Finland crossbows were always spanned with the *krihake*, spanning strap and spanning belt. Simple *krihake* were in use in southern Norrland in the 15th century. An example with a horn sheave is known from Ångermanland (25:81). Since at

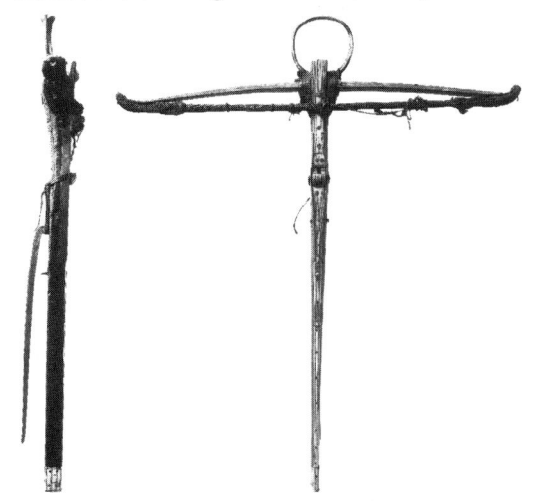

Figure 57 Crossbow with steel bow, spanned by a Samson belt. Northern Swedish or Finnish. 17th to 18th centuries. Livrustkammaren [Royal Armoury], Stockholm.

---

130 Alm perhaps underestimated the skill required to produce a large piece of spring steel, the two arms of which were of exactly the same strength. Anyone who has tried to make springs will tell of the many problems involved.

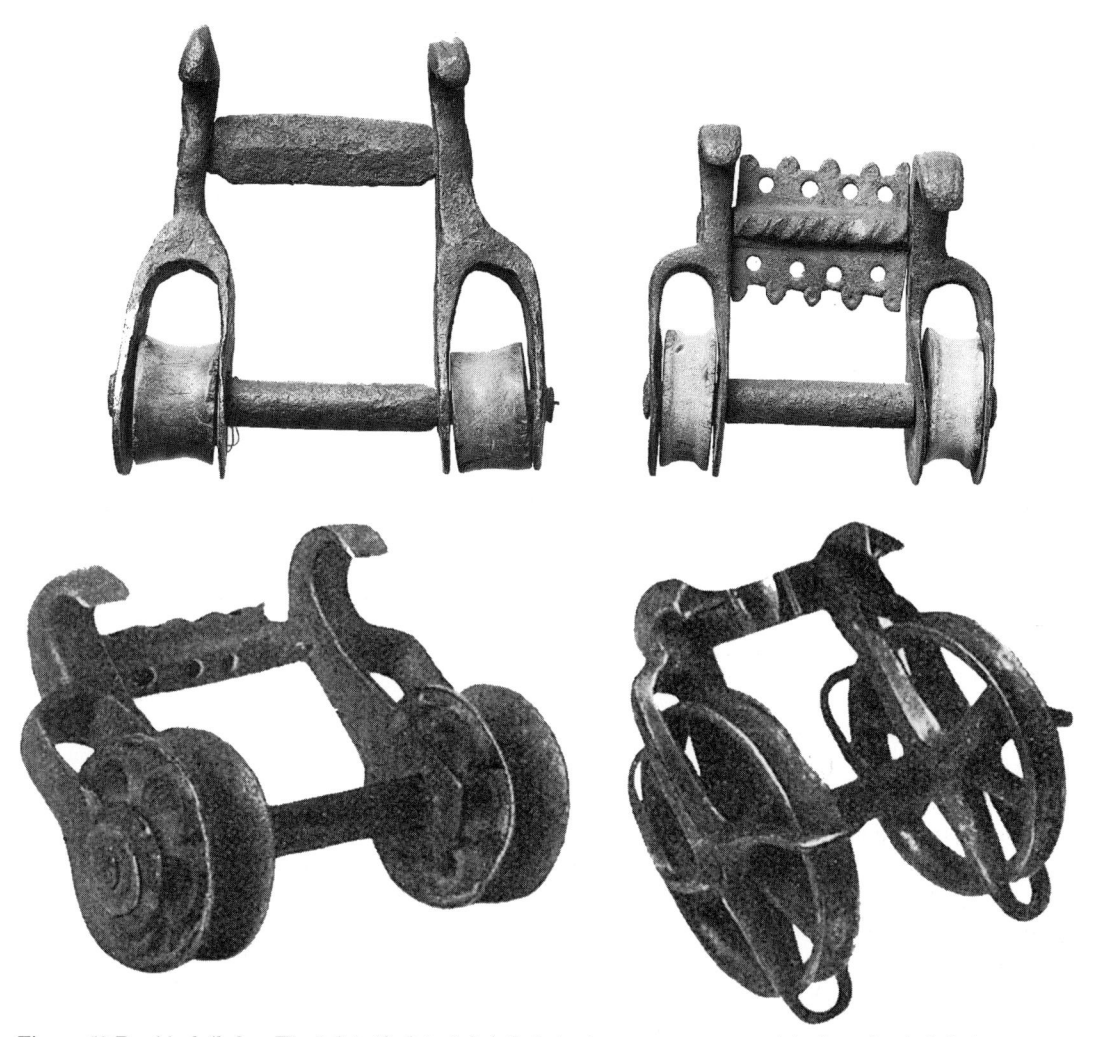

Figure 58 Double *krihakar*. The left half of the left *krihake* in the upper row was originally a simple *krihake*, to which the right half was added later. Ångermanland. The other *krihake* in the upper row also has horn pulley-wheels and may date from the close of the Middle Ages. Of the two *krihakar* in the lower row, the one on the left dates to the 16th century and the one on the right is of western European type, but may originally have formed part of an English windlass. Livrustkammaren [Royal Armoury], Stockholm.

least the 16th century, the *krihake* were always double, with pulleys first of horn and later of iron. Sometimes the iron pulleys have the shape of small wheels, as on the English windlasses. The *krihake* appears in the regions mentioned above in a number variants, a sign of local and temporal fashions.

Pistons, the bolts with blunted heads, appear in a number of variants (figure 59). They are usually 40-60cm long. The head most frequently has a diameter of about 5cm. On a number of Finnish types the head has a tenon projecting sometimes more, sometimes less far from its centre. The fletching is of exceptional length, and consists of either two or three flights. A thin layer of pitch is applied over the bindings. The end of the shaft, designed to lie between the fingers of the nut, is flat, and on later types is expanded so that its height is a good deal greater than its breadth.

Strangely, no information is available about hunting with the crossbow in Sweden in the

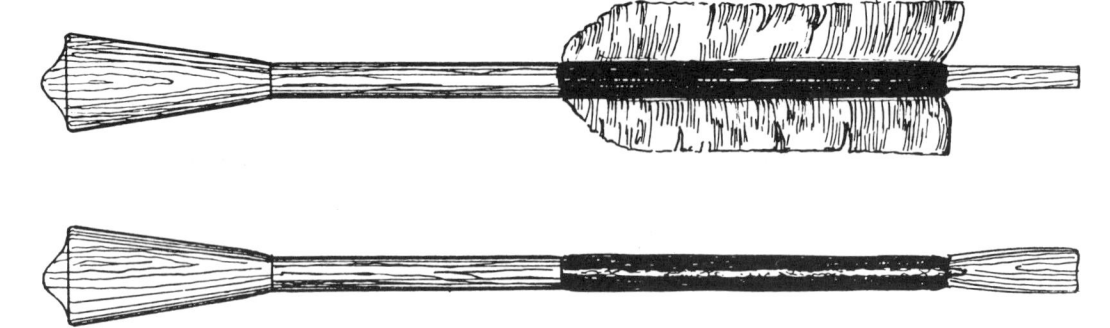

Figure 59 'Piston' for squirrel hunting. From northern Finland, length 53.4cm., weight 115g. Livrustkammaren [Royal Armoury], Stockholm.

16th century. Presumably, it was so commonplace at that time that no one took the trouble to make a record of it. On the other hand, there is a great deal of evidence about hunting with such weapons in the 18th century. At that time it was apparently usual to train dogs as bolt-retrievers. The evidence from many areas, Lycksele for example, suggests that the squirrel hunter was accompanied by a boy, the *'bolt-boy'* who was to watch where the bolts fell (31:154).[131]

Linnaeus saw the crossbow in use in Västerbotten in 1732. Curiously, he calls it the *'Norwegian crossbow'*, perhaps because the steel bow was imported from Norway, and he says that the bow was 2½ft long and that its breadth was two thumbs (inches) at the middle and one thumb at each end. The string was of hemp thread as thick as a finger, wrythen, or twisted, *'and laid about a little with the same sort of thread, particularly at the middle, where the nut was to contact it'*. The tiller was of wood, 2½ft long, and was decorated with inlaid bone. According to his drawing the pulleys of the *krihake* had the form of four-spoked wheels. Unfortunately, he does not say what name the West Bothnians had for the *krihake*. Regarding the *'piston'*, Linnaeus remarks that it was 1½ft long, was turned out of birch and that it had three fletchings of capercaillie feathers (69:167-168). Linnaeus says that these crossbows shot splendidly: *'I saw with astonishment how the piston knocked down a little twig I had set up 30 paces in front of the marksman'*, and *'they shoot squirrels very neatly at 20 to 30 paces, just the same as with a gun'* (69:167).

Castrén wrote of northern Finland in 1754:
'Squirrels are normally shot here from the middle of October throughout the whole winter by every man that has a steel bow and one or two good squirrel-hounds. A man who shoots squirrels always has with him a boy who is called Kolckamies whose function is to watch where the bolts or arrow fall and then to bring them to the marksman. With plenty of game and with good hounds a man can shoot between ten and fifteen or perhaps more squirrels a day'[132] (23:58).

Castrén also says that when a bear was scared out of its hibernating den it was pursued with hounds and was *'brought down either with a gun or with a bow or with a spear'* (23:60).

In the years 1798-9 J. Acerbi travelled in Sweden and Finland. The diary of his travels has been published in English, German and French. On the subject of squirrel hunting in northern Finland he says that the animal was hunted in the spring, by which he evidently means the winter, using crossbows and blunt-headed wooden bolts:
'They do not aim by holding the crossbow tiller

---

131 Lycksele is a town in southern Lapland, some 80 miles north west of Umeå.
132 The modern Finnish spelling of *Kolckanies* is *Kolkkammies*. It means literally 'piston bolt man'.

Figure 60 Finnish squirrel hunters, 1799. After Acerbi and Retzius.

next to their eye, but instead they hold the crossbow in front of them at the level of their belly, and despite this they rarely or never miss their target. When the bolt has fallen to the ground, they have it looked for at once, because it is too valuable to them for it to be left lying around' (1:290).

In the English edition of Acerbi's book there is a picture of a couple of Finnish squirrel hunters (figure 60). One of the hunters is shown spanning his crossbow. The spanning mechanism is, however, misunderstood. The hunter who is shooting is holding his crossbow

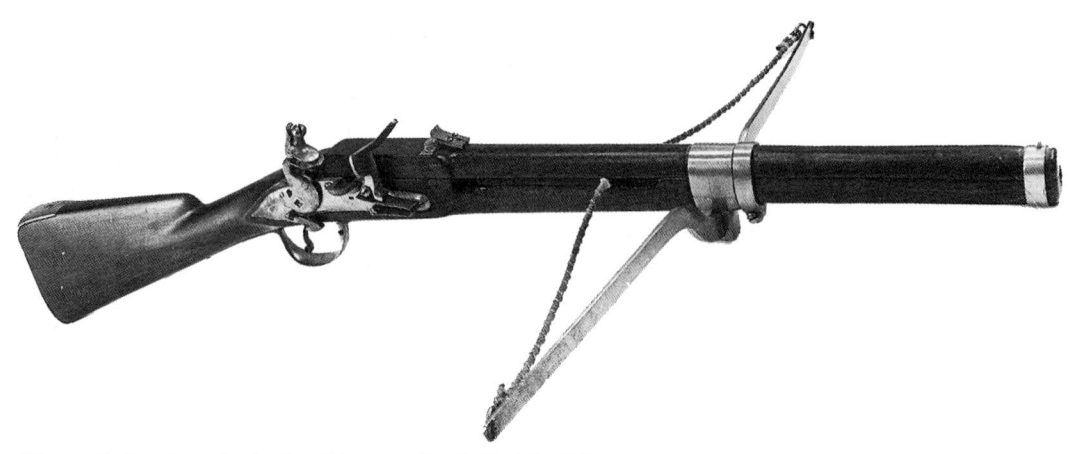

Figure 61 Crossbow for the Swedish navy. First half of the 19th century. Armémuseet [Army Museum], Stockholm.

in front of him with both hands and is shooting obliquely upward, exactly as Olaus Magnus showed the Swedish marksmen, shooting Dale bolts against hostile cavalry (figure 32).

When Professor Gustaf Retzius was in northern Tavastland in 1873 he met some old men who had hunted with crossbows in their youth.[133] In Pielavesi parish he was fortunate enough to buy five crossbow tillers. By then the bows had long since been refashioned into sickles for cutting grain, and the like (80:44). Retzius also bought a bolt which was 54cm long and had a head 10cm long with a tip of 5cm. As usual the threads that held the fletching in place were coated with pitch (80:46). Retzius says that the double *krihake* that was common in these areas of Finland was made either of iron or or brass (80:46).

I was told by the late Curator, Mr Keyland that he had proof that the crossbow was in use in some parts of southern Swedish Lapland as late as about 1815.

In the first half of the 19th century crossbows were used in, or at least were designed for, the Swedish navy. In the Armémuseum [Army Museum], Stockholm, and in the Statens Sjöhistoriska Museum [National Maritime Museum], Stockholm, there are some peculiar crossbows. The specimen in the Armémuseum (figure 61), has a tiller 102cm long, with an ordinary musket butt and a covered bolt-groove. The actual bolt-groove is 48cm long. The forward part of the tiller and the cover of the bolt-groove are held together by three iron bands. The one nearest the fore-end has a foresight on its upper side; and the one farthest back, which is 3.6cm broad, has on its underside a strong iron shoe with a rectangular hole in it to take the bow. This latter is held in place by a strong iron screw entering from behind. Towards the back of the barrel there is a folding sight which appears to have been taken from a model 1855 musket. On the right side of the tiller there is a flintlock which, like the sideplate, the swivel and the buttplate, has been taken from a model 1725 musket. The flintlock and the release mechanism for the string are linked together so that when fired the mainspring releases the simple apparatus that carries the string out of the spanning notch. The steel bow is 72cm long, and at the centre is 2.3cm broad and 0.8cm thick. The weapon weighs 3.2kg (AM 4000). In the Statens Sjöhistoriska Museum [National Maritime Museum], there are two more similar crossbows, from which the flintlocks and the barrel bands have been removed.

In the same collection there is also another

---

133 Tavastland was the Swedish name for what is now the Finnish province of Hämeen lääni, which begins some twenty miles north of Helsinki.

crossbow of a different type, which has been altered from a model 1747 musket. The barrel has a groove machined into it for the string. The fore-sight is of the same type as that on the model 1854 musket. Here, too, the flintlock and the crossbow lock seem to be linked together, but the crossbow lock has a claw release and appears to be spanned by a part projecting from the rear end or the barrel. A bolt accompanies the weapon. The head has a cylindrical body and a conical point. Behind the head the shaft has a cylindrical swelling fitted to the bore of the barrel. In its other proportions the shaft is heavy, and at its rear end it broadens out into a pair of flights.

Officially the flintlock was withdrawn from naval service in 1833, and therefore these weapons should date from before then. It is hard to say for what these crossbows were actually used. It has been suggested that they were intended for shooting fire-arrows, in which case the flintlock must have been intended for lighting the fire-arrows. However, there is a distance of 11.5cm between the pan and the beginning of the bolt-groove, and the bolt-groove does not seem to have any connection whatever with the pan. I suspect that these weapons were used, or were designed to be used, during the first stages of teaching servicemen the art of marksmanship, the idea being to accustom personnel to the flash of the powder in the pan, which is very disturbing to people who are not used to it.

At one place in Norway crossbows and a type of poisoned bolt were still in use for whale hunting around 1900. The whale concerned was the Bay whale, the smallest of all the whalebone whales, which can grow to a maximum of 10m in length but which is usually smaller.[134] These whales often entered the fjords for food. On the island of Store Sartor in the approaches to Bergen are three deep fjords with very narrow entrances. When whales entered these fjords the entrances were closed with nets made out of bast rope, from which the whales could not escape.[135] There they tried to shepherd the whales into some minor inlet. If they succeeded, the inlet was barred with ordinary herring nets, of which the whales were very frightened, even though such a net does not constitute the least hindrance to them when aroused (21:138f).

When a whale had been penned, the hunters started to shoot bolts at it. The crossbows used for this purpose have been described in the first chapter. The very thick bolt-shaft (figure 3), is of pine or spruce. The tang of an iron head about 12-15cm long was pushed into a hole in the centre of the shaft's forward end. For the majority of its length the point has a circular cross-section and a diameter of 5-6mm, but at the tip it has a triangular blade with either rounded or sharp angles at the base. The centre of gravity lies approximately at the middle of the bolt (21:145,148).

The spanning mechanism and the spanning process have been described in the first chapter. Once the crossbow was spanned, and the bolt was put in position, the shooter assumed a shooting position - similar to the Finnish squirrel-hunters in Acerbi's time, and like that sometimes used in Sweden towards the end of the Middle Ages when shooting was done in an obliquely upward direction - in which the crossbow is held at the level of the hips with both hands and in such fashion that the front of the tiller was higher than the back (21:146).

If the bolt missed, the shaft made it float and it was fished out as fast as possible. However, if it hit, the shaft detached as it met the whale's skin and the point continued on alone through the blubber and into the flesh. A number of

---

134 The Bay Whale is usually now known in English as the Minke Whale, although it has also been referred to as the Lesser Rorqual or Pike Whale. There are two types of whales, toothed whales (sub-order *Odontoceti*) and whalebone or baleen whales (suborder *Cylsticeti*), which instead of teeth have a lattice work of triangular baleen plates with hairy fringes which trap the Krill on which the whales feed.
135 Bast or bass is the inner bark of the lime tree, but the term is often more loosely used for other flexible, fibrous barks.

Figure 62 Danish trap crossbow of 1687. After Tantzern.

bolts were shot into each whale, which does not seem to have felt the wounds to any great extent. The bolts did not kill the whale immediately. After being shot the whale was left alone, with people keeping watch to see it did not break the net. After a few days or so the whale became very ill and could then easily be harpooned and killed. What made the bolts dangerous was that the points were rusty and covered with bacteria, which brought about a gangrene-like blood poisoning. When these whales were cut up, it was usually found that the flesh was inflamed around some of the bolt heads and that around one of them the flesh was rotten and stinking *'over the diameter of a small keg'* as an 18th century author says. The bolt that had made such a wound was called the 'death bolt'. It was carefully removed, but was not cleaned. The affected flesh area was also cut out and fresh bolt heads were thrust into it, and allowed to remain there for at least 12 hours. When they were subsequently removed, they were not cleaned, and were then considered to be death bolts. That they had indeed become for, through this treatment, they had become coated with pathogenic bacteria. When they were later shot into a whale they occasioned a similar appalling wound. The bolts were carefully preserved in special leather cases. Because of the treatment described above the heads were usually very rusty, but they were not cleaned, although if a tip became dull it was sharpened (21:147f).

The meat from a whale killed in this way could be used as food without the least danger, and it was a welcome change from the diet of fish (21:153). It is difficult to say how old this method of poisoning bolts was, but it gives the impression of being a Stone Age technique. In the past this method was presumably used at many points along the Norwegian coast, and perhaps in other countries as well. After 1900, the Krag-Jörgensen rifle took the place of the crossbow and poisoned arrows in whale-hunting. With a 6.5mm rifle of this sort one can shoot a whale dead.

The trap crossbow was used in Scandinavia just as it was in Germany, usually for foxes. In Count Gabriel Oxenstierna's armoury at Rosersberg in 1678 there were *four steel bows*

Figure 63 Crossbow trap for foxes. Used up to the middle of the 19th century. Småland. Nordiska Museet [Nordic Museum], Stockholm.

to catch foxes with' and three *skäktas* (74:26), which evidently went with these steel bows.[136] In Scandinavia, and presumably also in Germany, a special type of bolt was used with the trap crossbow. In his hunting manual, printed in Copenhagen in 1686 Tantzern mentions a trap crossbow (*Selbschuss*), (figure 62), when discussing hunting foxes, badgers and martens. As you can see, this is an ordinary crossbow that is spanned with a cranequin. However, the bolt is very remarkable, with its 15cm broad, crescent-shaped head (114:114, 118).

Crossbows were used in Småland as late as the middle of the 19th century as traps for foxes (figure 63). They were mainly of the same design as the German trap crossbow for lynxes that was described in Chapter 5. The majority of these Småland 'fox-bows' seem to have been altered from ordinary crossbows. The tiller [of the example illustrated] has been cut off 12.15cm behind the lock, although some seem to have been re-tillered at the time of conversion. The recess for the nut is usually simply cut out of the actual wood of the tiller, but occasionally the front of the recess is shod with iron or horn. The release mechanism lies on the right side of the tiller. The string sometimes consists of iron chain, in which cases the nut, too, is customarily of iron. The bow is held fast in the tiller by means of iron bridles, which are fastened in place with wooden wedges. A cord, the other end of which was tied to the bait, was attached to the trigger mechanism.

These fox bows were spanned with

---

136 Rosersberg Castle is situated some 20 miles north west of Stockholm on the main highway to Uppsala.

Figure 64 'Fox knife' bolts for crossbow traps. Småland. The larger one is 30.2cm. broad, 27cm. long, and weighs 220g. The smaller one is 16.5cm. broad and 27.5cm. long. Nordiska Museet [Nordic Museum], Stockholm.

cranequins, some of which date from as early as the 16th and 17th centuries, while others have been made specially for their present purpose, presumably during the 18th century or at the start of the 19th century. On the majority of these cranequins the loop that was placed over the tiller for spanning was of iron.

The heavy bolts that were used with these crossbows were called 'knives' or 'fox knives' (figure 64), and had evidently evolved from bolts of the type illustrated in figure 62. Fox knives have crescent-shaped heads 12.5-30cm broad. These latter often have sockets, but others, particularly the largest ones, have tangs. The shaft is usually quite thick, and square in section. Underneath and towards the rear, the shaft is cut to the form of the nut's fingers. Between the notches a sort of keel is formed which is pressed between the fingers of the nut so that the rear end of the shaft comes to rest against the spanned string.

The fox knives appear to have been intended to shoot off a fox's leg and in this way to make it difficult or impossible for him to run off (60:20). Such trap crossbows are also believed to have been used in Norway, for fox knives are to be found in the Sandvig Collection at Lillehammer.[137]

---

[137] Lillehammer is the main town of the county of Oppland in south east Norway.

8 Pellet Crossbows

In eastern Asia, India and certain areas of South America a type of bow is found that is not intended to shoot arrows, but rather to shoot pebbles or balls of fired clay. These pellet bows usually have two strings held apart by one or two small wooden stretchers. At their middle the strings are joined together by a small piece of leather or cloth, which constitutes the cup of a sling (13:67).[138]

In eastern Asia the pellet bow has evolved into a pellet crossbow. Such weapons are used, for instance, among the Karen tribes of Burma and Siam (29:91), and throughout China, which is probably the country of their origin.

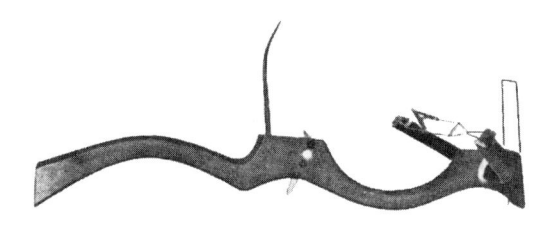

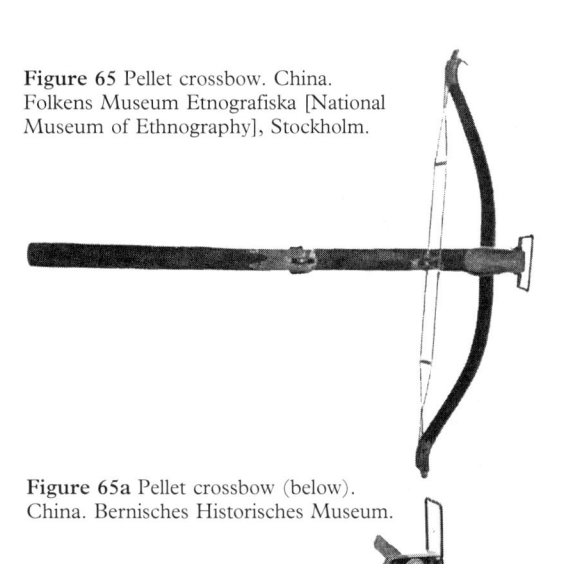

The Chinese pellet crossbow (figure 65), has a tiller of wood which is down-curved between the lock and the bow. The lock is of horn or bone and consists of only two parts, a claw and a trigger. To shoot, one must draw the claw and the string back a little before the string can be released. On the tiller ahead of the bow there is a little iron frame about 5cm high, across which a thread is stretched horizontally. At the centre of this thread a bead is fixed, which serves as a fore-sight. Behind the lock there is a tall, narrow tongue of wood or horn pierced with small sighting holes at various heights for different ranges. The bow is of horn. The string is double and the two cords are held at a suitable distance from each other by means of a pair of delicate wooden pins. At the centre there is a basket-like sling-cup made of cord, to the back of which is attached a strong loop for hooking over the claw. Chinese pellet crossbows are usually very small.

It was probably the Portuguese who brought Chinese pellet crossbows to Europe. In the middle of the 16th century weapons of this sort began to appear here, apparently at first in southern Europe. The oldest type of pellet crossbow is the Italian one (figure 66), which appears to have developed around 1550 (19:419). These pellet crossbows remind one a great deal of the Chinese ones, but they are usually larger. The lock is of a different design, and the bow is always of steel. Behind the lock the tiller is straight, and the butt usually swells into a knob.

Instead of the little iron frame that sits on the

Figure 65 Pellet crossbow. China. Folkens Museum Etnografiska [National Museum of Ethnography], Stockholm.

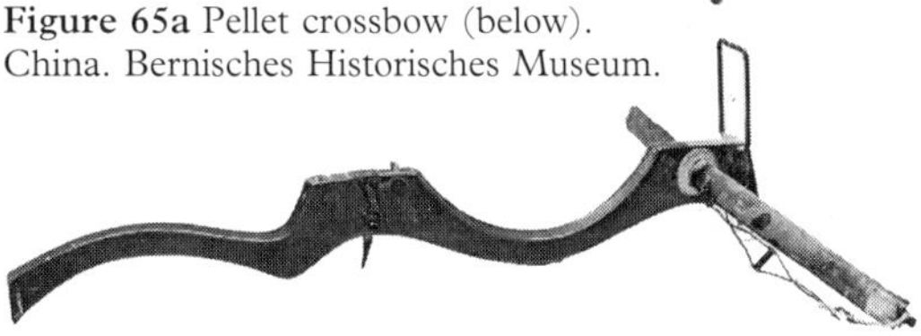

Figure 65a Pellet crossbow (below). China. Bernisches Historisches Museum.

---

[138] Pellet or stone bows were also used in Europe from at least the 14th century. See H L Blackmore *Hunting Weapons*, London, 1971, pp.164-9.

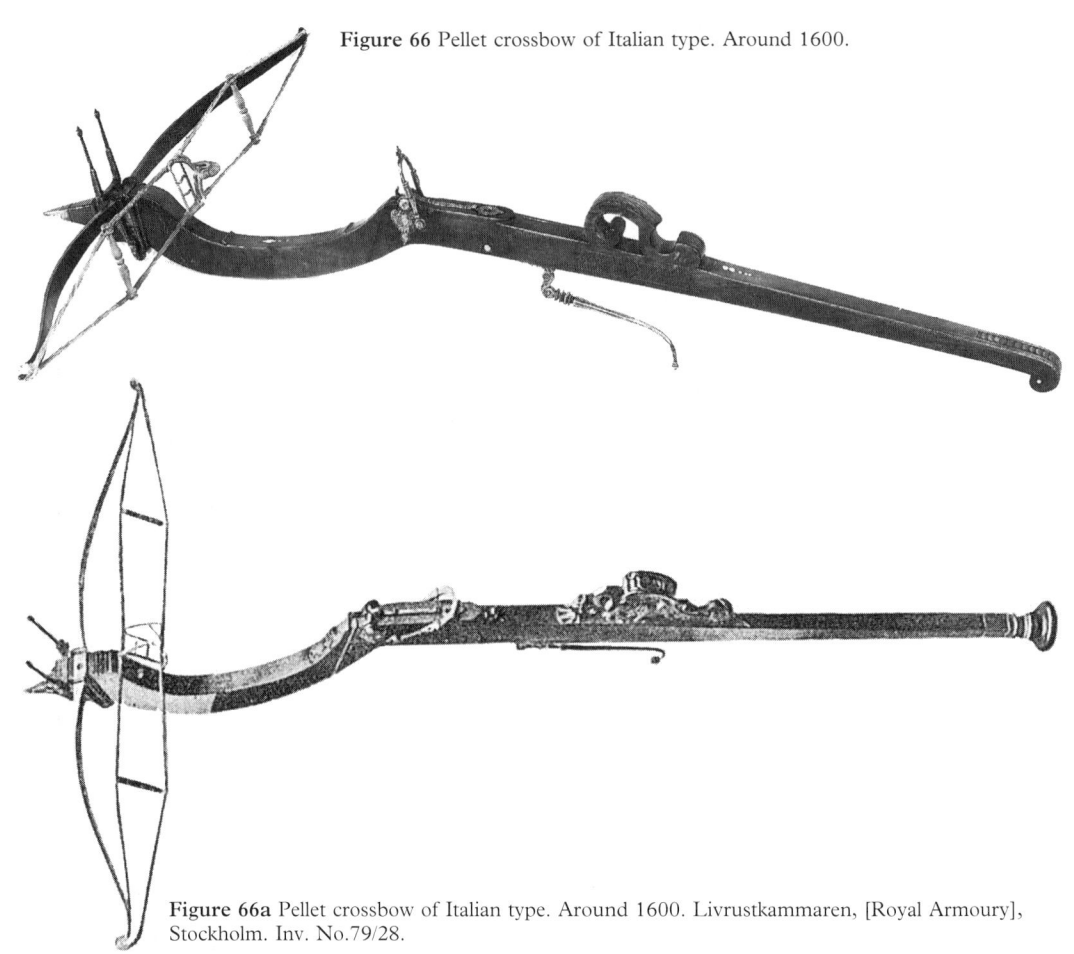

Figure 66 Pellet crossbow of Italian type. Around 1600.

Figure 66a Pellet crossbow of Italian type. Around 1600. Livrustkammaren, [Royal Armoury], Stockholm. Inv. No.79/28.

fore-end of the tiller of the Chinese pellet crossbows, the Italian ones have a small iron post on either side of the tiller fastened to the wood and rising about 5-8cm above the top of the tiller. These iron posts are joined by a thread which has a bead at its centre, to serve as a fore-sight. Behind the lock there is a folding frame, which has a sighting notch on its bowed upper side. The steel bow is fastened to the stock by means of iron bridles. The double string is made in the same way as those on Chinese crossbows. The pins that hold the two strings at the proper distance from each other have a notch for the cord at each end and are secured to the two strings by means of lashings. A crossbow of this type in Consul General Jean Jansson's collection was dated 1613. The length of the tiller was 75.5cm. The bow was 59cm long, and at the middle it was 1.15cm thick and 2cm broad. It weighed 1.38kg. These crossbows were usually spanned by hand. However, sometimes a sort of grapple equipped with hooks was used to afford a better grip in spanning, the hooks being securely engaged with the string (19:419). Pellet crossbows of this type were much used in the second half of the 16th century for hunting small birds and doves, particularly in Italy and Spain, but also in France and Germany. They appear to have

Figure 67 Pellet crossbow of German type. Made primarily of steel. End of the 16th century. Livrustkammaren [Royal Armoury], Stockholm.

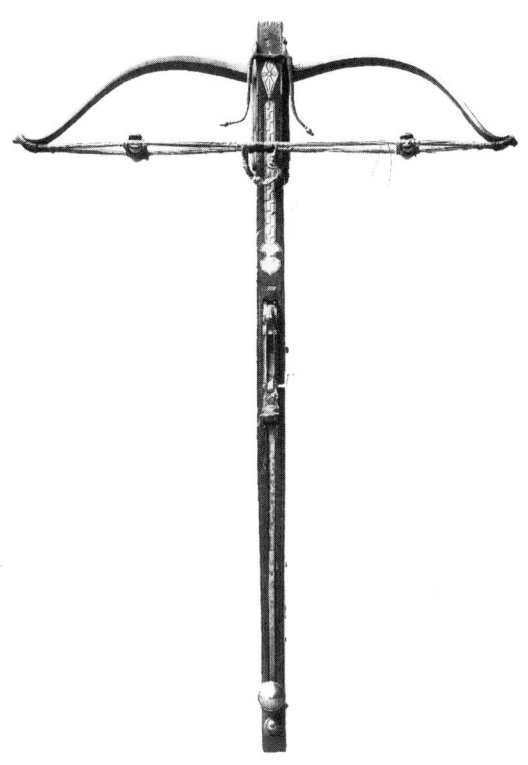

Figure 68 Pellet crossbow. Spanish or Dutch type. Beginning of 17th century. Livrustkammaren [Royal Armoury], Stockholm.

should properly be regarded as German, but such weapons do also seem to have been manufactured in northern Italy.

On the top of the tiller of these crossbows is a fork consisting of two parallel arms. Between the lines of the fork a thread is stretched, at the middle of which is positioned the bead that serves as a fore-sight. The lock is virtually identical to the English lock illustrated in figure 71, and like that one it is enclosed within a lock-housing 11-12 cm long, which is made up of two sturdy lockplates, one on either side. The lock is released by a hair-trigger recessed into the tiller, which is cocked by pressing a button on the underside of the tiller ahead of the trigger-guard which is quite long. The spanning mechanism is fixed to the tiller, and consists of a simple lever and the lock-housing to which this lever is fastened. The lever and the lock-housing are joined together by a pin around which the lock can pivot to a limited extent. This pin retains the high, folding back-sight which is pierced with a number of apertures. The bow is of steel, and passes through an opening towards the front of the tiller, where it is held fast by iron wedges. The string is just like that of the previous type.

gone out of use by the middle of the 17th century (19:419-421, 75:157).

Another European type of pellet crossbow was made entirely of steel, except for the butt which was of wood (figure 67). This type

The method of spanning is as shown in figure 70. When the lever is not in use, its rear end is secured to the tiller by means of locking

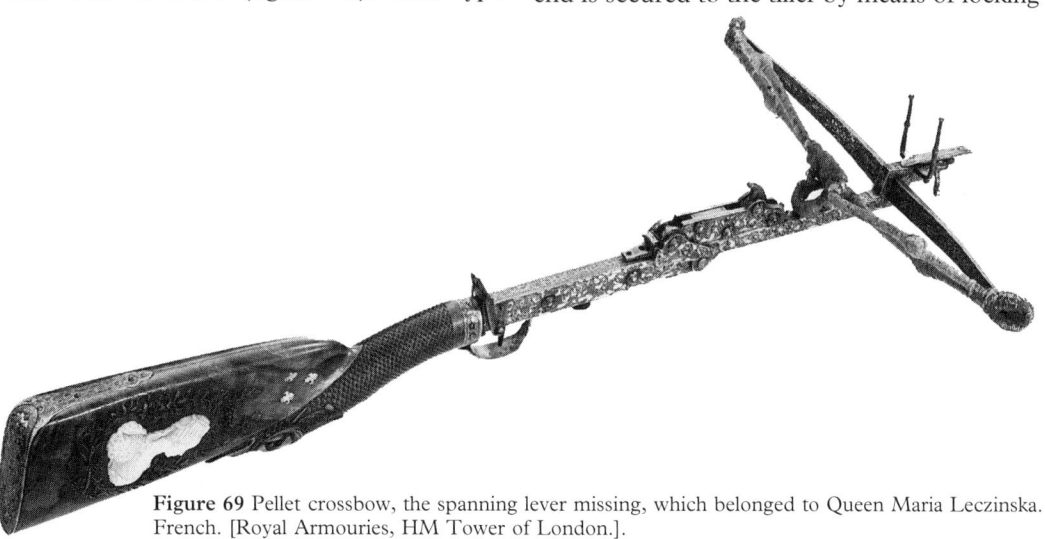

Figure 69 Pellet crossbow, the spanning lever missing, which belonged to Queen Maria Leczinska. French. [Royal Armouries, HM Tower of London.].

95

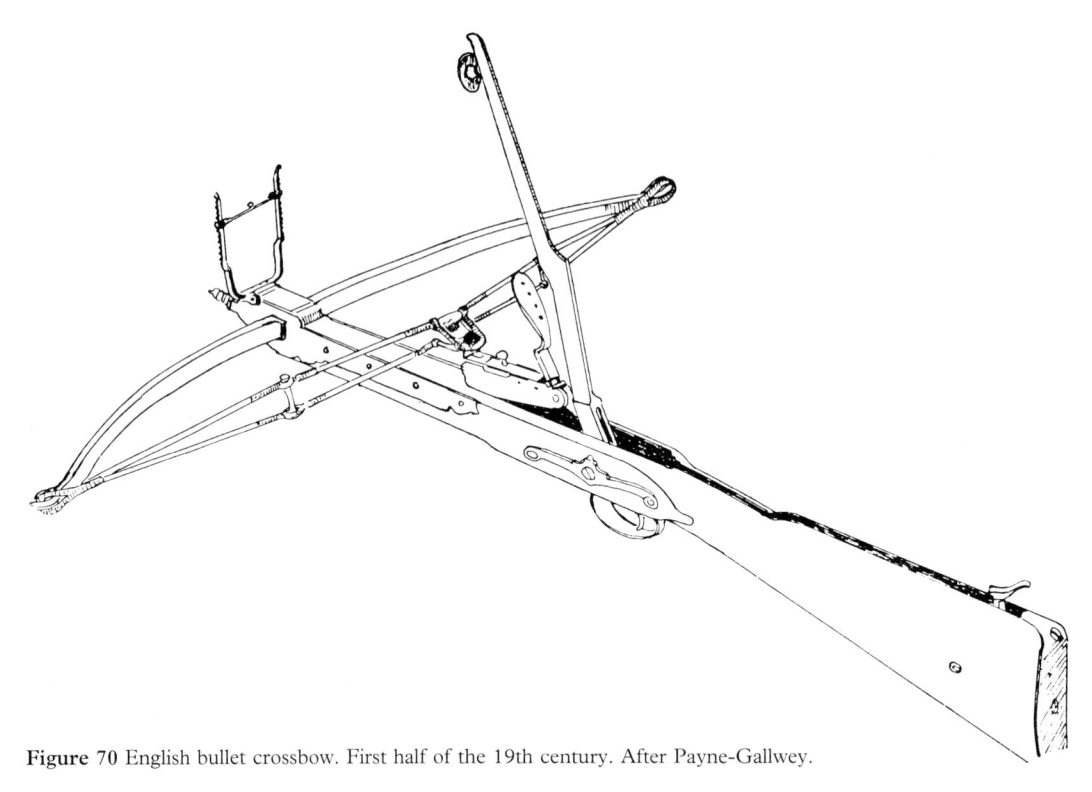

Figure 70 English bullet crossbow. First half of the 19th century. After Payne-Gallwey.

devices of different designs.

These crossbows were very small. When measured one of them was found to have a tiller 62cm long, and a steel bow 44.8cm long, 1.9cm broad and 1.9cm thick at the middle. It weighed 1.75kg (AM 3998). Larger crossbows of this type are found, but also ones that are even smaller. They were evidently more effective than the Italian pellet crossbows, and they seem to have made their appearance in the latter half of the 16th century. Chinese repeating crossbows have a spanning device that is constructed on exactly the same principle as that of the European crossbows described above. There are, in fact, repeating pellet crossbows in China.

In his hunting manual, printed in 1754, the German Döbel remarks that the pellet crossbow (*Palester*) is used for hunting small birds (28:224f), but he does not offer any more detailed information about the weapon. In Spain and the Netherlands pellet crossbows with wooden tillers were used from the first half of the 17th century onwards (figure 68). The sights, the lock and the spanning mechanism were like those on the German pellet crossbows described above. When the lever was not in use it lay in a recess in the upper side of the tiller behind the lock.

Pellet crossbows evidently continued in use in France during the 18th century. In Lieutenant Colonel Th. Jakobsson's arms collection there was a pellet crossbow which, according to its coat of arms and cipher, belonged to Louis XV's queen, Maria Leczinska (figure 69). It is very similar to the German pellet crossbows made of steel which have been described above, but the butt has been given the shape popular for the shoulder stocks of 18th century hunting rifles.[139]

---

139 The bow was acquired by the Royal Armouries in 1952 from the collection of William Randolph Hearst (Inv.No. XI.93). It was made by Trincks of Strasbourg, who was active between 1714 and 1732, and it therefore could have belonged to Maria Leczinska. The shoulder butt is of the type popular in France in the early years of the l9th century, and bears a mother of pearl cameo bust of Queen Marie Antoinette, and an engraved silver plaque of Diana.

Figure 71 Lock for an English bullet crossbow. After Payne-Gallwey.

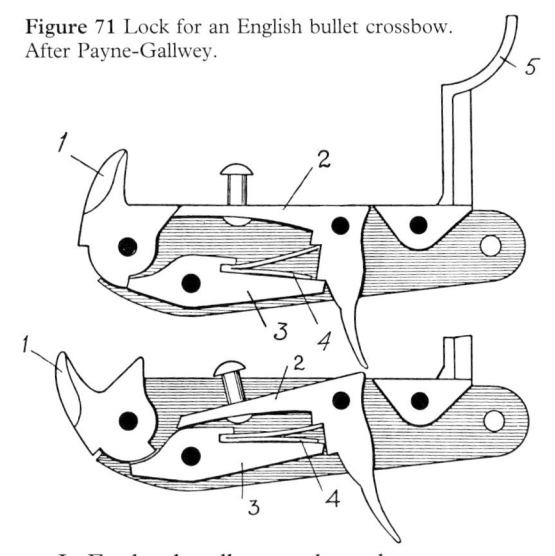

In England, pellet crossbows became popular again around 1800.[140] The English variant (figure 71), is much like the Hispano-Dutch type just described, but the tiller has a standard gun-butt. The lock, the double string and the spanning mechanism were similar to those on the German pellet crossbows of the latter half of the 16th century, and the same is true of most of the sighting devices. The iron fork, between the arms of which ran a thread with a bead attached to it, had on each of its tines a number of notches, so that the bead which served as a fore-sight could be set in different positions. However, this design had also appeared earlier as well.

The bow was fastened to the tiller with iron bridles. The lock (figure 71), consists of the claw 1, the upper lever 2, the lower lever 3 and the lever spring 4. The lock is operated by a trigger set in the tiller, which operates against the downward projecting part of the upper lever by means of an arm located upon its upper side.

These crossbows usually had steel bows 76cm long that were 19mm broad and 15.8cm thick at the middle (75:179). Payne-Gallwey reported that at a range of 18-22m he could hit a playing card eight times out of ten with this sort of crossbow. He also says that usually lead bullets weighing 14g were used as projectiles, and that if such a bullet was shot at an iron disc at a range of 1m the bullet was flattened into a hemisphere (75:199).

The English pellet crossbows were in use until around 1840-50, mainly for hunting rooks and rabbits. Subsequently they have been used by poachers, mostly against pheasants.

In Belgium, pellet crossbows of an entirely different type are still in use to a limited extent. They are heavy, greatly resemble the Belgian crossbows intended for target shooting with bolts which have already been described, and are equipped with the same sort of spanning-lever. However, the Belgian pellet crossbows have covered tillers, which constitute a sort of barrel of 16mm calibre. The bows are usually 79cm long, and are 3cm broad and 1.6cm thick at the middle. A lead pellet shot from such a crossbow has a range of up to 340m These weapons are used mainly for a kind of popinjay shooting, where one has to shoot down small wooden birds from the top of a mast 30m tall (75:210-221).

The pellet crossbow has apparently never been common in Sweden. In 1561 Eric XIV received *'a crossbow to shoot lead with'* from Karl Holgersson Gera (2:225). In 1609 Jakob the stockmaker received four dalers for making a tiller for a pellet crossbow (*en Kuulbåga låda*) for Duke Karl Filip (73:285).[141]

---

140 More recent research has shown that pellet crossbows were used in England continuously from the 16th century. During the second half of the 18th century bullet bows made especially in Lancashire, East Anglia and London, became increasingly popular. It is these bullet bows which Alm describes.
141 Duke Karl Filip (1601-1625) was the son of King Charles IX of Sweden, and brother of King Gustavus Adolphus. In 1611 he was offered the crown of Russia, but it went instead to Michael Romanov who immediately declared war on Sweden. Duke Karl Filip formally renounced his claims to the Russian throne in 1614.

9 References

Editor's Abbreviations

Z.H.W.: Zeitschrift für Historische Waffenkunde (Journal for the Study of Historic Weapons)

Z.H.W.K.: Zeitschrift für Historische Waffen und Kostümkunde (Journal for the Study of Historic Weapons and Costumes)

1. Acerbi, J. *Travels through Sweden, Finland, and Lapland*, vol.1, London, 1802.
2. Adlersparre, G. 'Afhandling om svenska krigsmaktens och krigskonstens tillstånd ifrån Gustaf I's död till Gustaf Adolphs tillträde till regeringen,' *Kongl. Vitterhets, Historie, och Antiquitets Academiens handlingar*, Del 3, Stockholm, 1793. (Dissertation on the state of military power and military science from the death of Gustaf I to the accession to the throne of Gustavus Adolphus,' *Proceedings of the Royal Academy of Science, History, and Antiquity*, vol.3, Stockholm, 1793).
3. Ahnlund, H. 'Gillena och Gustav Vasa'. *Hävd och Hembygd*, 1923. (The Guilds and Gustav Vasa, magazine '*Hävd och Hembygd*', 1923).
4.-6. Allen, C.F. *De tre nordiske Rigers Historie under Hans, Christiern den Anden, Frederik den Første, Gustav Vasa, og Grevefeiden*, Del 1, 2 og 3:2., Copenhagen, 1865-68. (*The History of the Three Northern Kingdoms under Hans, Christian II, Frederick I, and during the Feud of the Counts*, vols.1, 2, 3.2, Copenhagen, 1865-68.)
7. Alm, J. *Eldhandvapen*, Del 1, Stockholm, 1933 (*Hand Firearms*, vol.1, Stockholm, 1933).
8. Alm, J. 'Vapenhistoriska notiser från det senmedeltida Stockholm och Arboga' *Vaabenhistoriske Aarbøger*, 4 c, Copenhagen, 1945. ('Notes on the history of weapons from late medieval Stockholm and Arboga', '*Vaabenhistoriske Aarbøger*', vol.4c, Copenhagen, 1945.)
9. *Anna Komnena's Alexiade*, ovs. af O. A. Hovgaard, Del 2, Copenhagen, 1882, ('O A Hovgaard [trans] *The Alexiad of Anna Comnena*, vol.2, Copenhagen, 1882).
10-12. *Arboga stads tänkebok*, Utgiven av Erik Noreen och Torsten Wennerström, Del 1-2, Uppsala 1935-40. (Erik Noreen and Torsten Wennerström (ed), *The Memorandum Books of the City of Arboga*. vols.1-2, Upsala 1935-40.)
13. Archer (Alm J), *Bågar och bågskytte*, Stockholm, 1930. (Bows and Bowmen, Stockholm, 1930.)
14. Arvidsson, A.J. *Handlingar till upplysning om Finlands hävder*, Del 4, Stockholm, 1851. (*Papers discussing the Traditions of Finland*, vol.4, Stockholm, 1851.)
15. Balfour, H. 'The Origin of the West African Crossbow', *Journal of the African Society*, vol.8, London, 1909.
16. Berger, A. *Die Jagd aller Völker im Wandel der Zeit*, Berlin, 1928. (The Hunt among all peoples at all times, Berlin, 1928.)
17. Bergström, O. *Bidrag till Kongl. Uplands regementes historia*, Stockholm, 1882. (*Contribution to the history of the Royal Upland Regiment*, Stockholm, 1882.)
18. Blom, O. 'Blidemestere, Ballistarii og Vaerkmestere i Kjøbenhavn ca 1375-1550, *Historisk Tidsskrift*, Femte Raekke, Femte Bind, Copenhagen, 1885. (Catapult makers, Crossbow-makers, and Master Artisans in Copenhagen about 1375-1550, journal '*Historisk Tidsskrift*,' fifth series, vol.5, Copenhagen, 1885.)

19. Boeheim, W. *Handbuch der Waffenkunde*, Leipzig, 1890. (*Handbook for the Study of Weapons*, Leipzig, 1890.)

20. Brahe, P. d. ä (the Elder). *Fortsättning av Peder Svarts krönika. Utgiven av Otto Ahnlund*, Lund, 1896-1897. (*Continuation of Peder Svart's Chronicle, edited by Otto Ahnlund*, Lund, 1896-1897.)

21. Brunchorst, J. 'Hvalfangst med bue og pil', *Naturen. Illustreret maanedsskrift for populaer naturvidenskab*. Udg. af Bergens Museum, Tredie Raekke, 3:die Aargang, Bergen, 1899. ('Whale-hunting with bow and arrow', 'Nature,' Illustrated Monthly for Popular Natural Science, published by the Bergen Museum, third series, Year 3, Bergen, 1899.)

22. Buschan, G. *Illustrierte Völkerkunde*, Del 3, Stuttgart, 1922. (*Illustrated Ethnology*, vol.3, Stuttgart, 1922.)

23. Castrén, E. *Beskrifning öfver Cajaneborgs län*, Åbo, 1754. (*Description of the Province of Kajaani*, Abo [Turku], 1754.)

24. Cederström, R. 'Beobachtungen an älteren Bolzen', *Z.H.W.*, Del 6, Dresden, 1912-1914. ('Observations on Early Bolts', *Z.H.W.*, vol.6, Dresden, 1912-1914.)

25. Cederström, R. 'En tidig armborstspännare'. *Livrustkammaren, Band 2, Stockholm, 1941*. ('An early crossbow spanner', *Livrustkammaren*, vol.2, 1941.)

26. Dihle, H. 'Das Kriegstagebuch eines deutschen Landsknechts um die Wende des 15 Jahrhunderts' *Z.H.W.K.*, Band 3, Berlin, 1929-1931. 'War diary of a German Landsknecht at the turn of the 15th century', *Z.H.W.K.*, vol.3 , Berlin, 1929-1930.)

27. Dreyer, W. *Naturfolkens liv*, Stockholm, 1899. (*The Life of the Children of Nature*, Stockholm, 1899.)

28. Döbel, H.W. *Jaegerpractica*, Del 2, Leipzig, 1754. (*The Hunter's Practice*, vol.2, Leipzig, 1754.)

29. Egerton, W. *Description of Indian and Oriental Armour*, London, 1896.

30. Ehrenthal, M.von. *Führer durch die Königliche Gewehr Galerie zu Dresden*, Dresden, 1900. (*Guide to the Royal Arms Gallery at Dresden*, Dresden, 1900.)

31. Ekman, S. *Norrlands jakt och fiske*, Upsala, 1910. (*Hunting and Fishing in Norrland*, Uppsala, 1910.)

32. Engel, B. 'Nachrichten über Waffen aus dem Tresslerbuche des Deutschen Ordens', *Z.H.W.*, Band I, Dresden, 1897-1899. ('Data on Weapons from the Record Books of the Teutonic Order', *Z.H.W.*, vol.1, Dresden, 1897-1899.)

33. Falkman, L.B. *Om mått och vikt i Sverige*, Del 1, Stockholm, 1884. (*On Weights and Measures in Sweden*, vol.1, Stockholm, 1884.)

34. Gessler, E.A. 'Eine Armbrust aus der Westschweiz', *Z.H.W.*, Band 8, Dresden, 1918-1920. ('A crossbow from western Switzerland', *Z.H.W.*, vol.8, Dresden, 1918-1920.)

35. Grancsay, S.V. *The Bashford Dean Collection of Arms and Armor in the Metropolitan Museum of Art*, Portland, Maine, 1933.

36. Gripsholm. 'Utdrag ur ett inventarium öfver 'Archelidt' på Gripsholms slott anno 1554', *Handlingar rörande Skandinaviens historia*, Del 37, Stockholm, 1856. ('Extract from an inventory of 'Archelidt' at Gripsholm castle in 1554', *Transactions Concerning the History of Scandinavia*, vol.37, Stockholm, 1856.)

37. *Gustav Vasas registratur 1526*, Stockholm, 1865. (*Gustav Vasa's Registry, 1526*, Stockholm, 1865.)

38. *Gustav Vasas registratur 1530*, Stockholm, 1877. (*Gustav Vasa's Registry, 1530*, Stockholm, 1877.)

39. *Gustav Vasas registratur 1531*, Stockholm, 1877. (*Gustav Vasa's Registry, 1531*, Stockholm, 1877.)

40. *Gustav Vasas registratur 1532*, Stockholm, 1883. (*Gustav Vasa's Registry, 1532*, Stockholm, 1883.)

41. *Gustav Vasas registratur 1540*, Stock-

holm, 1891. (*Gustav Vasa's Registry, 1540*, Stockholm, 1891.)
42. *Gustav Vasas registratur 1543*, Stockholm, 1893. (*Gustav Vasa's Registry, 1543*, Stockholm, 1893.)
43. *Gustav Vasas registratur 1544*, Stockholm, 1895. (*Gustav Vasa's Registry, 1544*, Stockholm, 1895.)
44. *Gustav Vasas registratur 1545*, Stockholm, 1896. (*Gustav Vasa's Registry, 1545*, Stockholm, 1896.)
45. *Gustav Vasas registratur 1546*, Stockholm, 1900. (*Gustav Vasa's Registry, 1546*, Stockholm, 1900.)
46. *Gustav Vasas registratur 1547*, Stockholm, 1900. (*Gustav Vasa's Registry, 1547*, Stockholm, 1900.)
47. *Helsingelagen*. Utgiven av C.J. Schlyter, Lund, 1838. (*The Helsinge Law*, edited by C.J. Schlyter, Lund 1838.)
48. Hewitt, J. *Ancient Armour and Weapons in Europe*, Oxford and London, 1855.
49. Hildebrand, H. *Sveriges medeltid*, Del 2, Stockholm, 1898. (*Sweden in the Middle Ages*, vol.2, Stockholm, 1898.)
50. Holm, T. *Kungl. Västmanlands regementes historia*, Del 1, Stockholm, 1929. (History of the Royal Västmanland Regiment, vol.1, Stockholm, 1929.)
51. Horwitz, H.T. 'Die Armbrust in Ostasien', *Z.H.W.*, Band 7, Dresden, 1915-1917. ('The Crossbow in East Asia', *Z.H.W.*, vol.7, Dresden, 1915-1917.)
52. Hyltén-Cavalius, G.O. *Wärend och wirdarne*, Del 2, Stockholm, 1868. (*Värend and its people*, vol.2, Stockholm, 1868.)
53. Jansson, S.O. 'Mått och vikt i Sverige till 1500-talets mitt', *Nordisk Kultur*, Del 30, Stockholm, 1930. ('Weights and Measures in Sweden to the middle of the 16th Century', journal '*Nordisk Kultur*,' part 30.)
54. *Jomsvikingarnas saga, Isländska sagor*, översatta av A.U. Bååth, ånyo utgivna av Erik Olsson, Stockholm, 1925. (*The Jomsvikingasaga, Icelandic sagas*, translated by A.U. Bååth, and edited by E. Olsson, Stockholm, 1925.)
55. *Jyske Lov . Danmarks gamle Landskabslove*, Bind 2, Copenhagen, 1926. (*The Laws of Jutland. Denmark's Old Regional Laws*, vol.2, Copenhagen, 1926.)
56. Jahns, M. *Handbuch einer Geschichte des Kriegeswesens von der Urzeit bis zur Renaissance*, Leipzig, 1880. (*Manual on the History of Warfare from the Earliest Times to the Renaissance*, Leipzig, 1880.)
57. *Jönköping stads tänkebok 1456-1548*, utgiven av Carl M. Kjellberg, Jönköping, 1919. (*Memorandum Books of the City of Jönköping, 1456-1548.*, edited by Col. M. Kjellberg, Jönköping, 1919.)
58. *Kalevala*, Översatt av K. Collan, Helsingfors, 1922. (*The Kalevala*, translated by K. Collan, Helsingfors, 1922.)
59. *Karlskrönikan. Svenska Medeltidens Rimkrönikor*, Del 2. Stockholm, 1866. (*The Karl Chronicle. The Swedish Medieval Rhymed Chronicles*, part 2, Stockholm, 1866.)
60. Keyland, N. *Jaktavdelningen i Nordiska Museet Vägledning*, Stockholm, 1911. (*Guide to The Hunting Section in the Nordic Museum*, Stockholm, 1911.)
61. Klemming, G.E. *Skrå-ordningar*, Stockholm, 1856. (*Guild Regulations*, Stockholm, 1856.)
62. *Kongespegelen*, ovs. ved K. Audne, Del 2, Oslo, 1914. (*The King's Mirror*, translated by K. Audne, vol.2, Oslo, 1914.)
63. Köhler, G. *Die Entwickelung des Kriegswesens und der Kriegführung in der Ritterzeit*, Del 3:1, Breslau, 1887. (*The Development of Warfare in the Knightly Period*, vol.3:1, Breslau, 1887.)
64. Landau, G. *Beiträge zur Geschichte der Jagd und der Falknerei in Deutschland*, Kassel, 1849. (*Contributions to the History of Hunting and Falconry in Germany*, Kassel, 1849.)
65. Lenk, T. 'En nederländsk målskjutningsbössa i Skoklostersamlingen', *Livrustkammaren*, Band 3, Stockholm, 1943-

45. ('A Dutch Target Gun in the Skokloster Collection', *Livrustkammaren*, vol.3, Stockholm, 1943-45.)

66. Lenk, T. *Flintlåset, Dess uppkomst och utveckling*, Stockholm, 1939. (*The Flintlock, its Origin and Development*, Stockholm, 1939.)

67. Lenk, T. 'Medeltidens skjutvapen', *Nordisk kultur*, 12 B, 'Vapen', Stockholm, 1943. ('Projectile weapons of the Middle Ages', '*Nordisk kultur*', 12 B, 'Weapons', Stockholm, 1943.)

68. Lindblom, G. *Jakt och fångstmetoder bland afrikanska folk*, Del 2., Stockholm 1926. (*Hunting and Methods of Catching Game among African Peoples*, vol.2, Stockholm, 1926.)

69. Linne, C. von. *Iter Lapponicum*, utgiven av Th. M. Fries, Uppsala, 1913. (Linnaeus, *Iter Lapponicum*, edited by Th. M. Fries, Uppsala, 1913.)

70.-72. Magnus, O. *Historia om de nordiska folken*, Del 2,3, och 4, Uppsala, 1912, 1916, och 1925. Compared at doubtful points with the Latin edition of 1555. (Magnus, O. *History of the Northern Peoples*, parts 2, 3, 4, Uppsala, 1912.)

73. Malmborg, A.G. och Meyerson, Å. *Stockholms Bössmakare*, Stockholm, 1936. (*The Gunsmiths of Stockholm*, Stockholm, 1936.)

74. Meyerson, Å. 'Greve Gabriel Gabrielssons Oxenstiernas rustkammare på Rosersberg', *Livrustkammaren*, Band 3, Stockholm, 1943. ('The Armoury of Count Gabriel Gabrielsson Oxenstierna at Rosersberg', *Livrustkammaren*, vol.3, Stockholm, 1943.)

75. Payne-Gallwey, R. *The Crossbow*, London, 1903.

76. Petri, G. *Kungliga första livgrenadjärregementets historia*, Del 1, Stockholm, 1926. (*History of the First Royal Grenadier Guard Regiment*, vol.I, Stockholm, 1926.)

77. Pihlström, A. *Kongliga Dalregementets historia*, Del 1, Stockholm, 1902. (*History of the Royal Regiment of Dalarna*, vol.1, Stockholm, 1902.)

78. Potier, O. 'Criminalistische Betrachtungen über der Genuesermesser', *Z.H.W.*, Band 1, Dresden, 1897-9. (Criminological Considerations regarding the Genoa Knife, *Z.H.W.*, vol.1, Dresden, 1897-9.)

79. Rathgen, B. *Das Geschütz im Mittelalter*, Berlin, 1928. (*Shooting in the Middle Ages*, Berlin, 1928.)

80. Retzius, G. *Finland i Nordiska Museet*, Stockholm, 1881. (*Finnish Objects in the Nordic Museum*, Stockholm, 1881.)

81. Rohde, F. 'Die Abzugsvorrichtung der frühen Armbrust und ihre Entwickelung', *Z.H.W.K.*, Band 4, Berlin 1932-4. ('The Release Device of the Early Crossbow and its Evolution', *Z.H.W.K.*, vol.4, Berlin 1932-4.)

82. Rohde, F. 'Ueber die Zusammensetzung der spätmittelalterlichen Armbrust'. *Z.H.W.K.*, Band 7, Berlin, 1940-2. ('Concerning the Way the Late Medieval Crossbow is Put Together', *Z.H.W.K.*, vol.7, Berlin, 1940-2.)

83. Rose, W. 'Ueber die Waffen der Kreuzfahrer', *Z.H.W.*, Band 9, Berlin, 1921-2. ('On the Arms of the Crusaders, *Z.H.W.*, vol.9, Berlin, 1921-2.)

84. Rose, W. 'König Johann der Blinde von Böhmen und der Schlacht bei Crezy'. *Z.H.W.*, Band 7, Dresden 1915-7. ('King John the Blind of Bohemia and the Battle of Crécy', *Z.H.W.*, vol.7, Dresden 1915-7.)

85. Rääf, L.T. *Samlingar och anteckningar till en beskrivning över Ydre härad*, Del 5, Norrköping, 1875. (*Gleanings and Notes for a Description of the Ydre District*, part 5, Norrköping, 1875.)

86. Sahlström, G. *Rumlaborg*, Huskvarna, 1934. (*The Rumlaborg Fortress*, Huskvarna, 1934.)

87. *Saxes Danasage*, oversat af Jörgen Olrik, Bind 4, Copenhagen, 1909. (*The History of

101

the Danes by Saxo Grammaticus, translated by Jörgen Olrik, Copenhagen, 1909.)
88. Schnittger, B, och Rydh, H. *Aranäs*, Stockholm, 1927. (*Aranäs*, Stockholm, 1927.)
89. Schröder, J.H. *Inventarium Curiae Tynnelsö, 1443*, Uppsala, 1839. (*The Inventory of the Curia of Tynnelsö, 1443*, Uppsala, 1839.)
90. Sixl, P. 'Entwicklung und Gebrauch der Handfeuerwaffen', *Z.H.W.*, Band 2, Dresden, 1900-2. ('The Evolution and Use of Hand Firearms', *Z.H.W.*, vol.2, Dresden, 1900-2.)
91. Sixl, P. 'Geschichte des Schiesswesens der Infanterie', *Z.H.W.*, Band 2, Dresden, 1900-2. ('History of Missile Weapons Used by Infantry', *Z.H.W.*, vol.2, Dresden,1900-2.)
92. Spak, F.A. *Bidrag til handskjutvapnens historia*, Stockholm, 1890. (*Contribution to the History of Portable Projectile Weapons*, Stockholm, 1890.)
93. *Stockholms Stads Jordebok, 1420-1474*, Stockholm, 1894. (*The Property Register of the City of Stockholm, 1420-1474*, Stockholm, 1894.)
94. *Stockholms Stads Skottebok, 1460-1468*, Stockholm, 1926. (*The Tax Registers of the City of Stockholm 1460-1468*, Stockholm, 1926.)
95. *Stockholms Stads Skottebok, 1501-1510*, Stockholm, 1889-1915. (*The Tax Registers of the City of Stockholm, 1501-1510*, Stockholm, 1889-1915.)
96. *Stockholms Stads Skottebok, 1516-1525*, Stockholm, 1935. (*The Tax Registers of the City of Stockholm, 1516-l525*, Stockholm, 1935.)
97. *Stockholms Stads Tänkebok, 1474-1483*, Stockholm, 1917. (*Memorandum Books of the City of Stockholm, 1474-1483*, Stockholm, 1917.)
98. *Stockholms Stads Tänkebok, 1483-1492*, Stockholm, 1924. (*Memorandum Books of the City of Stockholm, 1483-1492*, Stockholm, 1924.)

99. *Stockholms Stads Tänkebok, 1492-1500*, Stockholm, 1930. (*Memorandum Books of the City of Stockholm, 1492-1500*, Stockholm, 1930.)
100. *Stockholms Stads Tänkebok, 1504-1514*, Stockholm, 1931. (*Memorandum Books or the City of Stockholm, 1504-1514*, Stockholm, 1931.)
101. *Stockholms Stads Tänkebok, 1514-1520*, Stockholm, 1933. (*Memorandum Books of the City of Stockholm, 1514-1520*, Stockholm, 1933.)
102. *Stockholms Stads Tänkebok, 1553-1567*, Stockholm, 1939. (*Memorandum Books of the City of Stockholm, 1553-1567*, Stockholm, 1939.)
103. *Stockholms Stads Tänkebok, 1576-1578*, Stockholm, 1943. (*Memorandum Books of the City of Stockholm, 1576-1578*, Stockholm, 1943.)
104. *Stockholms Stads Tänkebok, 1578-1583*, Stockholm, 1945. (Memorandum Books of the City of Stockholm, 1578-1583, Stockholm, 1945.)
105. Styffe, C.G. 'Raven von Barnekows räkenskaper 1365-1367 över Nyköpings län', *Bidrag till Skandinaviens historia ur utländska arkiver*, Band 1, Stockholm, 1859. ('The Accounts of Raven von Barnekow, 1365-1367, for the Province of Nyköping', in '*Bidrag till Skaninaviens historia ur utländska arkiver*', vol.1, Stockholm, 1959.)
106. Styffe, C. G. 'Preussika hansestädernas beslut av den 12 Juli, 1395', *Bidrag till Skandinaviens historia ur utländska arkiver*, Band 2, Stockholm, 1864 ('Prussian Hanse Towns Decree of l2 July 1395, *Bidrag till Skandinaviens historia ur utländska arkiver*, vol.2, Stockholm, 1864.)
107. Svart, P. *Gustaf I's krönika*, Stockholm, 1912. (*The Chronicle of Gustav I*, Stockholm, 1912.)
108. *Sverres saga. Norges Kongesagaer*, 3, Kristiania, 1914. (*Sverri's saga. The sagas of the Norwegian kings*, 3, Kristiania, 1914.)
109. *Södermannalagen*, udgiven av C.J.

Schlyter, Lund 1844. (*The Laws of Södermanland*, edited by C.J. Schlyter, Lund 1844.)

110. Tessman, G. *Die Pangwe*, Leipzig, 1913. (*The Pangwe People*, Leipzig, 1913.)

111. Thordeman, B. *Alnsöhus*, Stockholm, 1920. (*The Castle of Alnsö*, Stockholm, 1920.)

112. Thordeman, B. *Armour from the Battle of Wisby, 1361.* vol I, Uppsala, 1939.

113. Troels-Lund, T.F. *Dagligt liv i Norden på 1500-talet*, Del 7, Stockholm, 1932. (*Daily life in the North in the 16th Century*, vol.7, Stockholm, 1932.)

114. Täntzern, J. *Der Dianen Hohe und Niedere Jagdgeheimnisz*, Del 2, Copenhagen, 1686. ('*The Higher and Lower Secrets of Diana's Hunt*', vol.2, Copenhagen, 1686.)

115.-116. Wegeli, R. 'Inventar der Waffensammlung des Bernischen Historischen Museums in Bern', Del 1 och 2, '*Jahrbuch des Bernischen Historischen Museums*, 18, 19, Bern, 1939-40. ('Inventory of the Arms Collection of the Bern Historical Museum at Bern, parts 1 and 2', in '*Jahrbuch des Bernischen Historischen Museums*, 18, 19, Bern, 1939-40.)

117. Wegeli, R. *Katalog der Waffensammlung im Zeughause zu Solothurn*, Solothurn, 1905. (*Catalogue of the Arms Collection in the Armoury at Solothurn*, Solothurn, 1905.)

118. Weltil, F.C. 'Alte Missiven 1444-1448', *Archiv des Historischen Vereins des Kantons Bern*, Band 19, Bern, 1912. ('Old Missives, 1444-1448', in '*Archiv des Historischen Vereins des Kantons Bern*', vol.19, Bern, 1912.)

119. Vigander, H., Seeberg, F.G. och Hoff, J.A. *Skyttersaken i Norge*, Frederikshald, 1927. (*Shooting in Norway*, Frederikshald, 1927.)

120. D Wiklund, K.B. *Några bilder från det gamla Lappland*, Uppsala, 1909. (*Some Scenes from Old Lapland*, Uppsala, 1909.)

121. Wilbur, C. M. 'The History of the Crossbow', *Annual Report of the Smithsonian Institution, 1936*, Washington, 1937.

122. Zornmuseet, Mora, *Tillfällig utställning nr. 1. Dalpilar, Armborst, och Stålbågar.* Katalog utarbetad av R. Cederström, Mora, 1941. (The Zorn Museum at Mora, *Occasional Exhibition No.1, Dale bolts, Crossbows, and Steel bows*, a Catalogue prepared by R. Cederström.)

123. *Historisk-topografiske Skrifter om Norge og Norske Landsdele, Forfattede i Norge i det 16. Aarhundrede*, ugfivne ved G. Storm, Christiania, 1895. (*Historical and Topographical Writings About Norway and the Norwegian Regions, Composed in Norway During the 16th Century*, edited by G. Storm, Christiania, 1895.)

124. Strömbäck, S. *Forskninger på platsen för det forna Nya Lödöse*, Goteborg, 1924. (*Studies on the Site of the Former Nya Lödöse*, Gothenburg, 1924.)

Bibliography

Editor's introduction
Since Alm wrote his *Survey* much has been published on crossbows. In 1979 a comprehensive bibliography of crossbow literature was produced by Sarah Barter, then, as now, the Librarian of the Royal Armouries. For the first impression of this volume Sarah Barter Bailey, as she now is, prepared a new bibliography which brought her previous work up to date. This has been further revised and additions have been made for the 1998 reprint of this volume.

Bibliographer's introduction
The original version of this bibliography was published as 'International Crossbow Bibliography' in *Rapport* 3–4 of the Institut Suisse d'Armes (Grandson, 1979). It acknowledged the compiler's indebtedness to F. Lake and H. Wright *A bibliography of Archery* (Manchester, 1974). It has since been twice updated, further expanded and corrected from the material in the Royal Armouries Library, but the debt to the original bibliographers remains.

The bibliography is arranged in the following order:

A. Works of a general character that include material on crossbows, their appearance, use and development.

B. General surveys of the development and use of the crossbow.

C. Studies of crossbows.
   1. in a particular country, region or period
   2. of a particular type or of a single bow

D. Bolts (quarrels) and bolt-heads.

E. Other parts and accessories: strings, mechanisms, cases, quivers, etc.

F. Use of the crossbow.
   1. in war
   2. in hunting
   3. for target shooting, including accounts of shooting societies

G. Construction of the crossbow.

H. Crossbows in particular collections.

International crossbow bibliography

A Works on the history of arms and armour

Anon. England in the Olden Time. No. 3: On Crossbows and Bows and Arrows. *Saturday Magazine* **XIV** 9 March 1839, pp. 93–4. Illustrated.

— De l'Art Militaire Chez les Arabes au Moyen Âge. *Journal Asiatique, series 4* (?Paris) **XII** 1848, pp. 193–237. Includes bows and crossbows.

d'Allemagne, H. R. *Sports et Jeux d'Adresse.* (Paris) 1904. Includes 'Jeu de l'Arc et de l'Arbalète', pp. 75–90. Illustrated.

Auceps Sporting Amusements for April 1861. *New Sporting Magazine* new series **XLI** April 1861, pp. 294–5. Mentions use of the bullet crossbow for shooting birds.

Basel – Museum für Völkerkunde *Technologie frühzeitlicher Waffen.* (Basel) 1963. 'Die Armbrust' p. 40. Illustrated.

Blackmore, H. L. *Hunting Weapons.* 1971. Chapter 5: Crossbows, pp. 172–215. Illustrated.

— *Les Plus Belles Armes de Chasse du Monde.* (Geneva) 1983. Includes a few decorated crossbows, illustrated.

Blaine, D. P. *Encyclopaedia of Rural Sports.* 1840. Includes 'The Modern Crossbow', 146 pp. Illustrated.

Blair, C. A Note on the Early History of the Wheellock. *Journal of the Arms and Armour Society* **III** 1961, pp. 221–56, plates 45–7, 69. Includes discussion of early combined crossbows/wheellock guns.

— *European and American Arms, c. 1000–1850.* 1962. Includes 'The Crossbow', pp. 34–8, plates 236–43.

Boeheim, W. Bogen und Armrust. Eine vergleichende Studie über Gebrauch und Wirkung alter Fernwaffen. *Zeitschrift für Historische Waffenkunde* (Dresden) **I** 1898, pp. 133–5; 161–4. Illustrated.

Brett, R. J. *A Pictorial and Descriptive Record of the Origin and Development of Arms and Armour.* 1894. Includes 'The Crossbow', pp. 108–10, and plates including pavises.

Brunner, C. *Die Verwundeten in den Kriegen der Alten Eidgenossen.* (Tübingen) 1903. Not seen, cited Harmuth 1971 (**F1**) as including discussion of wounds inflicted by crossbow bolts.

Brusewitz, G. *Hunting: Hunters, Game, Weapons and Hunting Methods from the Remote Past to the Present Day.* 1969, original Swedish edition 1967. Includes 'The Crossbow', pp. 74–5. Illustrated.

Buttin, C. Notes sur les Armures à l'Épreuve. *Revue Savoisienne.* (Annecy) 1901, reprinted separately, 100 pp. Illustrated. Discusses the use of crossbow bolts to test the effectiveness of armour. See also Buttin 1906 (**D**).

Cahen, C. Un Traité d'Armurerie Composé pour Saladin. *Bulletin d'Études Orientales* (Paris) **XII** 1948, pp. 103–65. Illustrated, includes crossbows.

Carre, J. B. L. *Panoplie, ou Réunion de Tout ce qui a Trait à la Guerre, Depuis l'Origine de la Nation Française jusqu'à nos Jours.* (Chalons-sur-Marne and Paris) 1797. Includes 'Armes offensives de Trait: Arbalètes, Traits, Crochet, Cranequin', pp. 265–9, plate VIII: 'Arcs à jalets...', pp. 270–9, plate IX.

Clephan, R. C. *The Defensive Armour and Weapons and Engines of War of Medieval Times and of the Renaissance.* 1900. Includes 'The Crossbow', pp. 183–5, illustrated.

— Notes on Roman and Medieval Military Engines, etc. *Archaeologia Aeliana* (Newcastle upon Tyne) New series, **XXIV** 1903, pp. 69–114. 'The crossbow', pp. 98–105.

— The Ordnance of the 14th and 15th Centuries. *Archaeological Journal* **LXVIII** 1911, pp. 49–138. Includes garrots shot from early cannon, pp. 65–7.

Couissin, P. Les Armes Romaines. *Essai sur les Origines et l'Evolution des Armes Individuelles du Legionnaire Romain.* (Paris) 1926. Includes 'L'arbalète' p. 487.

Cowper, H. S. *The Art of Attack, Being a Study in the Development of Weapons and Appliances of*

Offence from the Earliest Times to the Age of Gunpowder. (Ulverston, Lancashire) 1906. Includes 'The Bow and Crossbow', pp. 207–26; 'The Stone or Pellet Bow', pp. 227–8. Illustrated.

Daremberg, C. V. & **F. Saglio**, *Dictionnaire des Antiquités Grecques et Romaines.* 5 vols (Paris) 1873–1919. Includes 'Arcuballista', 'Manuballista', I, p. 388; 'Gastraphetes', II, p. 4459.

Dean, B. A Crusaders' Fortress in Palestine: a Report of Explorations Made by the Museum, 1926. *Bulletin of the Metropolitan Museum of Art* (New York) **September** 1927, Part II, pp. 46, text illustrated. Finds included crossbow nuts and bolt heads.

Demmin, A. *Die Kriegswaffen in Ihren Geschichtlichen Entwicklungen.* (Leipzig) 1893 (4th edn) and earlier editions which all include a chapter 'Die Armbrust' with illustrations. Translated into several languages, e.g., *An Illustrated History of Arms and Armour* 1877. Includes 'The Crossbow', pp. 473–84; 'Siege Engines', pp. 455–62.

Deraniyagala, P. E. P. Sinhala Weapons and Armour. *Journal of the Royal Asiatic Society, Ceylon Branch* (Colombo) **XXXV** 1942, pp. 97–142. Illustrated, includes bows, arrows and crossbows.

Dillon, H. A. L. (Viscount Dillon) Arms and Armour at Westminster, the Tower and Greenwich, 1547. *Archaeologia* **LI** 1888, pp. 219–80.

— On the Development of Gunlocks from Examples in the Tower of London. *Archaeological Journal* **L** 1893, pp. 115–31. Includes discussion of crossbow triggers, pp. 115–7.

— Armour notes. *Archaeological Journal* **LX** 1903, pp. 96–136. Includes discussion of double proof-mark on a helmet of Henry VIII's in the Tower of London, said to be proof against the giant crossbow.

Elmy, D. Ainu Archery. *Journal of the Society of Archer-Antiquaries* **21** 1978, pp. 21–4. Text illustrated.

Duchartre, P. L. *Histoire des Armes de Chasse et de leurs Emplois.* (Paris) 1955. Includes 'Les Arbalètes', pp. 107–24. Illustrated.

Feldhaus, F. M. *Die Technik.* (Munich) 1965 (Facsimile reprint). Includes 'Armbrust', etc.

Francis, P. H. *Mechanical Biology: Announcing the Discovery of Twelve Biological Rules or Laws,* nd. [1952]. Includes bows and crossbows, pp. 317–469, 482–6, 519–28. Illustrated.

Gardner, Robert Edward *Small Arms Makers: a directory of fabricators of firearms, edged weapons, crossbows and polearms.* (New York) about 1963. 378pp., illustrated.

Gessler E. A. *Die Trutzwaffen der Karolingerzeit vom VIII bis zum XI Jahrhunderts.* (Basel) 1908. Includes 'Bogen, Pfeil, Armbrust', pp. 13–22.

Giorgetti, G. *L'Arco, la Balestra e le Macchine Belliche.* Milan 1964. Includes 'La Balestra', pp. 27–89; 'Le macchine Belliche', pp. 90–134.

Grabowska, I. Trzy Okazy Renesansowej Broni Mysliwskiej z Deckoracja Trawiona. *Studia i Materialy do Dziejow Dawnego Uzbrojenia i Ubioru* Wojskowego(Cracow) **IV** 1969, pp. 25–53, plates X–XXVIII. English summary: 'Three Specimens of Hunting Weapons with Etched Decoration', pp. 50–3. Includes a cranequin.

Grohman, W. A. B. Ancient Weapons of the Chase. *Burlington Magazine* **Nov.** 1903, pp. 185–94; **Feb.** 1904, pp. 157–67. Illustrated, includes crossbows.

Hansard, G. A. *The Book of Archery.* 1840. Chapter VI: 'The Crossbow', pp. 226–59.

Heath, E. G. *The Grey Goose Wing.* 1971. Chapter 9: 'The Crossbow', pp. 284–314.

Heer, E. (ed.) *Der Neue Støckel, I–III.* (Schwäbisch Hall) 1978–89. Includes crossbow makers' marks.

Hendricks, C. S. Hand operated projectile weapons. *Bulletin of the American Society of Arms Collectors* (Dallas, TX) **14** 1966, pp. 28–33.

Hewitt, J. *Ancient Armour and Weapons in Europe (...) to the End of the 17th Century.* 1855, 3 vols. Includes crossbowmen, I, pp. 158–60, 201–4; II, pp. 29–35; III, pp. 520–3, 542–5, 606–7. Illustrated.

Higson, D. *Sea Fowl Shooting Sketches.* (Preston, Lancs) 1909. Includes 'Stringing the Crossbow', pp. 87–91.

Hoopes, T. T. Projectile Weapons – Bows and Crossbows. *Saint Louis Museum Bulletin* (Saint Louis, MO) **38** 1954, pp. 30–2

Jahns, M. *Entwicklungsgeschichte der Alten Trutzwaggen mit einem Anhang über die Feuerwaffen.* (Berlin) 1899. Includes 'Die Armbrust', pp. 333–8. Illustrated.

Kalervo, H. Zur Geschichte des Mittelalterliche Geschützwesens aus Orientalischen Quellen. *Studia Orientalia* (Helsinki) **IX** 3 1941, pp. 1–261.

Keller, M. L. *The Anglo-Saxon Weapon Names Treated Archaeologically and Etymologically.*

(Heidelberg) 1906. Includes 'Arcubalista', pp. 54–5.

Koehler, G. *Die Entwicklung des Kriegswesens und der Kriegführung in der Ritterzeit.* (Breslau) 1886–90, 3 vols. Includes 'Schuss und Wurfmaschinen' III, pp. 141–67: 'Die grossen Armbrüste' III, pp. 174–90.

Kyeser, C. *Bellifortis.* Umschrift und Ubersetzung, von Gotz Quarg. [Facsimile of Göttingen MSS. philos. 64]. (Düsseldorf) 1969. Ff. 774–9 show crossbows and accessories.

Lake, F. & H. Wright *Bibliography of Archery.* Simon Archery Foundation. (Manchester) 1974, 501 pp.

Laking, G. F. A. *Record of European Armour and Arms Through Seven Centuries.* 1920–2, 5 vols. Includes 'The Long Bow and the Crossbow' I, pp. 125–7. 'The Crossbow' III, pp. 127–44.

Latham, J. D. & W. F. Paterson *Saracen Archery.* 1970. An English version of a 14th-century Mameluke work on archery.

Leeuw, W. van der Looy De Boog. *Oude Kunst* (Haarlem) **V** 1920, pp. 53–62. Includes crossbows.

Lenk, Torsten *Medeltidens skjutvapen.* In *Vapen.* Ed. Bengt Thordeman. (Stockholm) 1943, pp. 134–156.

Lewerken, H. W. *Kombinations Waffen des 15–19 Jahrhunderts.* (Berlin) 1989. 308 pp. Includes combined crossbows and firearms.

London—British Museum *The Age of Dürer and Holbein: German Drawings 1400–1550.* 1988. Includes a reproduction of a 15th-century drawing of a man spanning a crossbow with a cranequin and a study of the crossbow.

Magne de Marolles, G. F. *La Chasse au Fusil.* (Paris) 1788. Includes the use of the pellet crossbow for shooting small birds.

Martínez de Espinar, A. (Que de el Arcabuz a su Majestad) *Arte de Ballistería y Montería (...).* (Madrid) 1644. In spite of this work's enthusiastic recommendation of the advantages of the silence or the crossbow as a hunting weapon, all the five engravings illustrate hunting with a gun.

Medvedev, A. F. *Rucnoe Metatel'noe Oruzie. Luk i Strely, Samostrel, VIII–XVI, VV.* (Missile Weapons: Bows, Arrows, Crossbows, 8th–16th Century). (Moscow) 1966.

Moloney, A. An Exhibition of Crossbows, Longbows, Quivers, etc. from the Yoruba Country (Nigeria) *Journal of the Anthropological Institute* **XIX** 1889, pp. 213–15.

Moseley, W. M. *An Essay on Archery, Describing the Practice of that Art in all Ages and Nations.* 1792. Chapter XII: On the Arbalest, pp. 281–310.

Norman, A. V. B. *Arms and Armour.* 1964. Includes bows and crossbows, pp. 119–28.

Oakeshott, R. E. *The Archaeology of Weapons.* 1960. Includes 'The Crossbow', pp. 298–300, illustrated.

Paterson, W. F. *Encyclopaedia of Archery.* 1984. Includes crossbow material, illustrated.

Peterson, H. L. *Arms and Armor in Colonial America, 1526–1783.* (Harrisburg, PA) 1956. Includes 'Crossbows', pp. 7–11.

Przyluski, J. L'Arc et l'Arbalète en Indo-Chine. *Anthropologie* (Paris) **XXXII** 1922, p. 283.

Rathgen, B. Die Feuer- und Fernwaffen in Naumburg von 1348–1449. *Naumburger Tageblatt* (Hamburg) **Sept.–Oct.** 1920, p. 48. Reprint, 1921.

Rose, W. Anna Komnena über die Bewaffnung der Kreuzfahrer. *Zeitschrift für Historische Waffen- und Kostümkunde* (Dresden) **IX** 1921–2, pp. 1–10.

Scott, B. W. The bow. *New Zealand Antique Arms Gazette* (Christchurch, NZ) June 1991, pp. 4–7.

Sewter, E. R. A. (trans.) *The Alexiad of Anna Comnena.* (Harmondsworth) 1969. Includes the famous description of the Frankish crossbow, pp. 316–7.

Stone, G. C. *A Glossary of the Construction Decoration and Use of Arms and Armor in all Countries and in All Times.* (Portland, ME) 1934. Includes 'Arbalest', 'Crossbow', etc.

Trench, C. C. *A History of Marksmanship.* 1972. Includes 'Archery', pp. 23–102, illustrated.

Viollet-le-Duc, E. E. *Dictionnaire Raisonné du Mobilier Français de l'Époque Carlovingienne à la Renaissance.* 6 vols (Paris) 1858–75. Includes 'Arbalète' V, pp. 20–38; 'Carreau' V, pp. 254–5.

Vladescu, C. M. Încercári Asupra Periodizárii si Tipologiei Armelor Albe Medievale Occidentale – Sec. XV–XVIII–II, Armuri si Arme de Aruncare la Distanta. (Resumé: Essais sur la Classification et la Typologie des Armes Blanches Médiévales; II, Armures et Armes de Trait) *Studii si Materiale de Muzeografie si Istorie Militara* (Bucharest) **II–III** 1970, pp. 108–39. Illustrated, includes crossbows.

Waring, T. A. *Treatise on Archery, or the Art of Shooting with the Long Bow (...) Likewise a Dissertation on the Steel Crossbow, with Directions for Using it.* 1814, and several later editions until 1847.

Whitelaw, Charles E. *Scottish Arms Makers: a biographical dictionary of makers of firearms, edged weapons and armour working in Scotland from the 15th century to 1870.* 1977. Ed. Sarah Barter, 339 pp.

Wilkinson, H. *Engines of War, or Historical and Experimental Observations on Ancient and Modern Warlike Machines and Implements.* 1841. Includes 'Balistae' and 'Catapultae', pp. 35–44; 'Crossbow', pp. 23–7.

Wood, J. *The Natural History of Man, Being an Account of the Manners and Customs of the Uncivilised Races of Men.* 1868. Includes the Fan crossbow, pp. 594–6: Chinese repeating crossbow, pp. 812–4 Illustrated.

B The Crossbow

Anon. The Arbalest or Crossbow. *Mirror of Literature, Amusement and Instruction* **19** 1832 17 March, pp. 161–4.

— The History of the Crossbow. *Notes and Queries* Series 4 **V** 1870 29 January, p. 120.

— Cruel Crossbows. *Notes and Queries* Series 8, **II** 1892 20 August, p. 147; 1 October, p. 273; 5 November, p. 377; **III** 1893 7 January, p. 17.

— The History of the Crossbow. *Notes and Queries* Series 10, **II** 1904 3 December, pp. 443–4.

Aston, C. *The Crossbow, a History. Guns and Weapons.* (Amersham, Bucks) Winter 1988, pp. 91–2. Illustrated.

Berg, S. & C. Schenk Die Armbrust. *Deutsches Waffen Journal* (Schwabisch Hall) August 1965, pp. 74–7 Illustrated.

Bosson, C. L'Arbalète. *Gazette des Armes* (Asnières) No. 27 May 1975, pp. 31–7. Illustrated.

Chernoff, A. M. Medieval Menace (the Crossbow). *Guns and Ammo* (Los Angeles) **XI**, March 1967, pp. 30–3, pp. 70–3. Illustrated.

Combs, Roger (ed.) *Crossbows.* (Northfield, IL) 1987. Illustrated.

Currelly, C. T. The Crossbow. *Toronto Royal Ontario Museum Bulletin* (Toronto) **X** 1931, pp. 11–12.

Curtis, H. History of the Crossbow. *Archery World* (Boyertown, PA) 15 December 1966, pp. 44–5; 16 February 1967, pp. 30–5; April 1967, pp. 44–5. Illustrated, with bibliography.

Gallwey, R. P. The Medieval Crossbow – Its Range by Comparison with the Longbow *Field* **99**, 4 January 1902, pp. 12–3.

— The Medieval Crossbow: with Some Notes on the Longbow. *Field* **99**, 1 February 1902, pp. 162–3.

— The Crossbow. Medieval and Modern. Military and Sporting, its Construction, History and Management, with a Treatise on the Balista and Catapult of the Ancients. 1903.

—*Appendix to the Book of the Crossbow and Ancient Projectile Engines.* 1907.

— Summary of the History. Construction and Effects in Warfare of the Projectile Throwing Engines of the Ancients. 1907.

Gilchrist, H. I. Ancient Crossbows as Works or Art. *Arts and Decoration* (New York), 19 June 1923, p. 11.

Gillet, A. V. *De Gescheidenis van de Kruisboog Geillustreerd door Postzegels.* (Antwerp) Nd, about 1970.

Harmuth, E. *Die Armbrust.* (Graz) 1975. An expanded and updated translation of Gallwey, 1903.

— *Die Armbrust: ein Handbuch.* (Graz) 1986.

Harris, P. V. From longbow to crossbow and back. *Journal of the Society of Archer-Antiquaries* 18 1975, pp. 9–12.

Hiestand, W. A. A One-time Rival of the Longbow (the Crossbow). *Sylvan Archer* (Alsea, OR) 6 April 1933, pp. 9–10.

Hoops, J. Die Armbrust im Frühmittelalter. *Wörter und Sachen* (Heidelberg) **III** 1912, pp. 65–8.

Horwitz, H T. Zur Entwicklungsgeschichte der Armbrust. *Zeitschrift für Historische Waffenkunde* (Dresden) **VIII** 1918–20, pp. 311–7; **IX** 1921, pp. 73, 114, 139.

Jenkinson, G. Four Centuries of the Crossbow. *Country Life* **CXXV** 16 April 1959, pp. 832–3.

Kalmar, J. A Közepkori Szamszerij [The Medieval Arbalest]. *Technikatörteniti Szemle* (Budapest) **III** 1964, pp. 97–127.

Kluge, F. Armbrust oder Armrust? *Zeitschrift für Historische Waffenkunde* (Dresden) **II** 1900–2, p. 233.

Koetschau, K. Zur Etymologie des Wortes Armbrust. *Zeitschrift für Historische Waffenkunde* (Dresden) **III** 1903–5, pp. 142–3.

Maglioli, V. La Balestra. *Armi Antiche* (Turin) **II** 1955, pp. 97–109.

Palotai, G. *Hungarian Crossbow Bibliography.* (Budapest) 1983. (Photocopied manuscript in the library of the Royal Armouries.)

Paterson, W. F. *A Guide to the Crossbow.* Edited with an introduction by Arthur G. Credland. A memorial volume published by the Society of Archer-Antiquaries, 1990.

Pieraccinia, S. La Balestra con Obbiettivita. *Diana Armi* (Florence) **November** 1986, pp. 70–3. Illustrated.

Potier, O. Armbrust oder Armrust? *Zeitschrift für Historische Waffenkunde* (Dresden) **II** 1900–2, pp. 322–3.

Rausing, G. Något om Armbrostets Äldre Historia. *Medd. Skånes Vapenhistoriska Förening* (Valmö) **93** November 1971, pp. 6–13.

Reid, W. The Crossbow – Ancient and Modern. Survival of a Medieval Weapon. *Shooting Times* **10 March** 1961, pp. 170–1, 286.

Scott, B. W. The bow. *New Zealand Antique Arms Gazette* **June** 1991, pp. 4–7

Spencer, C. L. Notes on the Crossbow. *Transactions of the Glasgow Archaeological Society* (Glasgow) New series **V** 3 1908, pp. 186–97.

Stefanska, Z. Genealogia Kuszy. *Muzealnictwo Wojskowe* (Warsaw) 1985, pp. 137–48. Illustrated, English title: The genealogy of the crossbow.

Wilbur, C. M. The History of the Crossbow Illustrated from Specimens in the United States National Museum. *Annual Report of the Smithsonian Institution* (Washington) 1936, pp. 427–38. Reprinted separately as Smithsonian Publication No. 3438 (Washington) 1937.

Cl The crossbow in a particular country, region or period

Alm, J. Europeiske Armborst – en Oversikt. *Vaabenhistoriske Aarbøger.* (Copenhagen) **Vb** 1947, pp. 1–255. Illustrations, bibliography, summary in German.

Balfour, H. The Origin of West African Crossbows. *Journal of the African Society* **VIII** 1909, pp. 337–56. Illustrated. Reprinted: *Annual Report of the Smithsonian Institution* (Washington) 1910, pp. 635–50, and separately.

Beveridge, H. Oriental Crossbows. *Asiatic Quarterly Review* 3rd series **XXXII** 1911, pp. 344–8.

Blackmore, H.L. An Archery Bill for Henry VIII, 1547. *Journal of the Society of Archer-Antiquaries* **32** 1989, pp. 5–8. Illustrated.

Burgesse, J.A. Montagnais Crossbows. *Beaver* (Winnipeg, Canada) December 1943, pp. 37–9.

Cotton, P. H. G. P. Notes on Crossbows and Arrows from French Equatorial Africa. *Man* **XXIX** 1929, pp. 1–3. Reprinted: *Journal of the Society of Archer-Antiquaries* (Colchester, Essex) **5** 1962, pp. 19–22.

Credland, A. G. The Crossbow in the Far North. *Journal of the Society of Archer-Antiquaries* **26** 1983, pp. 12–23. Illustrated.

— The Crossbow in Africa. *Journal of the Society of Archer-Antiquaries* **32** 1989, pp. 37–45. Illustrated.

—The crossbow and the law, from the dark ages to the present. *Journal of the Society of Archer-Antiquaries* **33** 1990, pp. 51–64.

Dite, J. Slowakische Jagdarmbrüste aus dem 17 Jahrhundert und ihre Besonderheiten. *Waffen- und Kostümkunde* (Munich) 3rd series **XVIII** 1 1976, pp. 40–52. Illustrated.

— Noch Einmal zu den Slowakischen Armbrüsten. *Waffen- und Kostümkunde* (Munich) 3rd series **XIX** 2 1977, pp. 160–1.

Dreher, A. Armbrust und Büsche in einer Alten Reichsstadt (Ravensburg). *Der Museumfreund* (? Ravensburg) **I** 1962, pp. 18–25.

Ekdahl, S. Die Armbrust im Deutschordensland Preussen zu beginn des 15. Jahrhunderts. *Fasciculi Archaeologicae Historicae* **5** 1992, pp. 17–48.

Elmy, D.J. The Han Dynasty Crossbow 206 BC–AD 221. *Journal of the Society of Archer-Antiquaries* (Swalecliff, Kent) **10** 1967, pp. 41–3. Illustrated.

Elmy, D.J. Notes on Eastern Crossbows. *Journal of the Society of Archer-Antiquaries* **18** 1975, pp. 25–7. Illustrated.

Forke, Dr The Chinese Crossbow, 1896. *Journal of the Society of Archer-Antiquaries* **29** 1986, pp. 28-33. Illustrated. A translation of an article in *Zeitschrift zur Ethnologie Verhandlunge* **XXVIII** 1896, pp. 272–6.

Gilbert, J.M. Crossbows on Pictish Stones. *Proceedings of the Society of Antiquaries of Scotland* (Edinburgh) **107** 1976, pp. 316–7.

Gunter, R. Die Schweizerische Armbrust. *Schuss und Waffe* (Neudamm) **I** 1908, pp. 265–68. Illustrated.

Harmuth, E. Die Armbrustbilder des Haimo von Auxerre. *Waffen- und Kostümkunde* (Munich) 3rd series, **12** 2 1970, pp. 127–130. Illustrated.

— Zur Einfussarmbrust der Hochgotik. *Waffen- und Kostümkunde* (Munich) 3rd series **XX** 1 1978, pp. 47–50. Illustrated.

— Concerning the One-foot Crossbow of the High Gothic. *Journal of the Society of Archer-Antiquaries* **28** 1985, pp. 9–12. Illustrated.

— Eine Arabische Armbrust. *Waffen- und Kostümkunde* (Munich) 3rd series **XXV** 2 1983, pp. 141–4. Illustrated.

Harrison, M. The Crossbow of the Baka

Pygmies. *Journal of the Society of Archer-Antiquaries* 31 1988, pp. 4–8.

Heer, E. Notes on the Crossbow in Switzerland. *Arms and Armor Annual* (Northfield) **I** 1973, pp. 56–65. Illustrated.

— Aus der Geschichte der Armbrust (in der Schweiz). *Die Armbrust-eine Sportwaffe, Herisau* (Verlap Schläpfer) 1976, pp. 3–30. Illustrated.

Horwitz, H.T. Die Armbrust in Ostasien. *Zeitschrift für Historische Waffenkunde* (Dresden) **VII** 1916, pp. 155–83, 170–4. Illustrated.

Hotchkis, J. Note on Crossbows in The Paston Letters, 1469. *Journal of the Society of Archer-Antiquaries* (Colchester, Essex) 4 1961, p. 36.

Jenkins, P. B. American Indian Crossbows. *Wisconsin Archaeologist* (Milwaukee) New series, **VIII** 1929, pp. 132–3.

Kirpichnikov, A. N. Kryuk dlya natyagivaniya samostrela (1200–1240gg). *Akademiya Nauk SSSR* (Moscow) 1971, pp. 100–102.

Knebel, K. Balistarii, Schützenmeister oder Armbrustmacher. Beitrag zur Geschichte des Älteren Handwerkes in Sachsen. *Freiberger Altertumsverein, Mittheilungen* (Freiberg) **XL** 1904, pp. 62–8.

Krohn, J.A.J. Krüppel. Eine Spätgotische Armbrust. *Deutsches Waffen Journal* (Schwäbisch Hall) February 1989, pp. 174–9. Illustrated.

Loewe, R. Jewish Evidence for the History of the Crossbow. *Le Juif au Miroir de l'Histoire* (Paris) 1985, pp. 87–107. Illustrated.

McEwen, E. Chinese and Korean Crossbow Bows. *Journal of the Society of Archer-Antiquaries* (Havant, Hants) 16 1973, pp. 32–3. Illustrated.

— The Bow of the Ox. *Journal of the Society of Archer-Antiquaries* 28 1985, pp. 15–21. Illustrated.

Morin, M. Una Balestrino del XVI Secolo. *Diana Armi* (Florence) June 1977, pp. 32–6. Illustrated.

Mortzsch, O. Preis von Pferden und Armbrust, 1440. *Zeitschrift für Historische Waffenkunde* (Dresden) **VIII** 1915, p. 112.

Mus, P. Les Balistes du Bayon. *Bulletin de l'École Française d'Extrême-Orient* (Hanoi, Vietnam) **XXIX** 1929, pp. 331–40 Illustrated.

Paskiewicz, M. Kusza i Topot. *Orzel Bialy* (Brussels/London) Nos 841, 2 1958, pp. 33–4. Translated into English: The Crossbow and the Axe. *Journal of the Arms and Armour Society* 3 12 1961, pp. 295–8.

Paterson, W.F. Maltese Crossbows. *British Archer* (Portsmouth, Hants) **VI** 1954–5, p. 152.

Skrivanic, G. Samostrel i Njegova Primena u XV Veku. *Vesnik* (Belgrade) **XI–XII** 1966, pp. 381–6. English summary: The Crossbow and its Application in the 15th Century.

Stefanska, Z. Genealogia kuszy. *Muzealnictwo Wojskowe* 1985, pp. 137–148.

Tang, M.C. Crossbows of the Formosan Aborigines and the Origin of the Crossbow. *K'ao ku jen lei hsuer k'an* (Taiwan University, Department of Archaeology and Anthropology Bulletin) (Taiwan) **XI** 1358, pp. 5–32, Illustrated. English summary p. 33.

Vitt, G. G. Jr Chu-ko-nu: the Manchurian repeating crossbow. *Journal of the Society of Archer-Antiquaries* 38 1995, pp. 10–16.

C2 Crossbows of a particular type

Anon. The Berkhamsted Bow. *Antiquaries Journal* **XI** 1931, p. 433.

Bishop, W. E. Webster Has it. *British Archer* (Portsmouth, Hants) **XIV** 1962–63, pp. 146–7. A discussion of the use of the word 'prod' for the bow of a crossbow.

Boothroyd, G. The Bullet Crossbow. *Shooting Times* (Windsor, Berks) 3–9 September 1981, p. 23. Illustrated.

Brown, R. C. Observations on the Berkhampstead Bow. *Journal of the Society of Archer-Antiquaries* 10 1967, pp. 12–17.

Brown, R.G. A Long-barrelled Cross-bow. *Journal of the Society of Archer-Antiquaries* 24 1981, pp. 31–4. Illustrated.

Buttin, C. L'Arc à Jalet. *Mémoires de la Société Savoisienne d'Histoire et d'Archéologie* (Chambéry) **LXIV** 1927, pp. 266–77. Illustrated. Pointing out that this was not a crossbow in the 15th century, although the term came to be used for one later.

Clephan, R.C. Note on an Arbalest or Windlass Crossbow. *Proceedings of the Society of Antiquaries of Newcastle-Upon-Tyne* (Newcastle Upon Tyne) Series 2 **X** 1902, p. 302.

Cosson, C.A. de. The Crossbow of Ulrich V, Count of Wurtemburg, 1460, with Remarks on its Construction. *Archaeologia* **LIII** 1893, pp. 445–63. Illustrated.

Credland, A. G. The Bullet Crossbow in Britain. *Journal of the Society of Archer-Antiquaries* (Swalecliffe, Kent) 15 1972, pp. 22–38. Illustrated.

— An Eighteenth-century Ratchet Crossbow. *Journal of the Arms and Armour Society* **VIII** 4 1975, pp. 48–51. Illustrated.

— Crossbow Guns and Musket Arrows, *Journal of the Society of Archer-Antiquaries* (Canterbury) **20** 1977, pp. 5–19. Illustrated.

— Daniel Higson and the Bullet Crossbow. *Journal of the Society of Archer-Antiquaries* **28** 1985, pp. 24–33. Illustrated.

— The Tiller Bow. *Journal of the Society of Archer-Antiquaries* **35** 1992, pp. 13–15. Illustrated.

Davenport, C. H. Crossbows. *Antiquary* **XXIII** 1891, pp. 149–52. Illustrated. Discusses three early 17th-century crossbows in the possession of Prebendary Baldwin-Childe of Kyre Park, Worcs.

Dean, B. A crossbow of Matthias Corvinus, 1489. *Bulletin of the Metropolitan Museum of Art* (New York) 1925, pp. 154–7. Illustrated.

Dresden—Historisches Museum Armbrust des Kurfürst Johann Friedrich des Grossmütigen von Sachsen um 1540. *Zeitschrift für Historische Waffen- und Kostümkunde* (Dresden) **IX** 1921, p. 174.

Dwyer, B. A Warring States Repeating Crossbow. *Journal of the Society of Archer-Antiquaries* **39** 1996, pp. 62–7.

Elmy, D. & N. Allen An Assassin's Crossbow. *Journal of the Society of Archer-Antiquaries* (Swalecliffe, Kent) **15** 1972, pp. 37–9. Illustrated.

— & **W. E. Flewett** The Assassin's Crossbow. *Journal of the Society of Archer-Antiquaries* **17** 1974, pp. 27–34. Illustrated.

— & **G. D. Gaunt** Chinese Stonebows. *Journal of the Society of Archer-Antiquaries* (Swalecliffe, Kent) **10** 1967, pp. 29–39. Illustrated.

Flewett, W. E. The Pistol Crossbow. *Journal of the Society of Archer-Antiquaries* (Canterbury) **20** 1977, pp. 22–4, illustrated.

— The Compound Crossbow. *Journal of the Society of Archer-Antiquaries* **35** 1992, pp. 6–7, illustrated.

— European Combination Weapons. *Journal of the Society of Archer-Antiquaries* **36** 1993, pp. 51–70, illustrated.

— The 'Assassin's Crossbow'? – a reassessment. *Journal of the Society of Archer-Antiquaries* **39** 1996, pp. 78–93, illustrated.

Frazer, W. Ancient Crossbow or Latch Obtained in Dublin During the Excavations in the Plunket Street Area, 1883. *Proceedings of the Irish Royal Academy* (Dublin) **XVI** 1884, p. 298.

Gates, E. Dainty but Deadly (A Pygmy Crossbow from Africa). *Bow and Arrow* (Covina, CA) **III** May 1965, pp. 62–5.

Gessler, E. A. Eine Armbrust aus der Westschweiz. *Zeitschrift für Historische Waffenkunde* (Dresden) **VIII** 1918–20, pp. 390–1. Illustrated.

Goddard, E. H. Notes on a Roman Crossbow, etc., Found at Southgrove Farm, Burbage. *Wiltshire Archaeological and Natural History Magazine* (Devizes, Wilts) **XVIII** 1894–95, p. 87.

Grancsay, S. V. An Enriched German Sporting Crossbow. *Los Angeles County Museum Bulletin* (Los Angeles) **VII** 1955, pp. 3–7.

Grosser, E. M. The Reconstruction of a Chou Dynasty Weapon. *Artibus Asiae* (Ascona, Switzerland/New York) **XXIII** 1960, pp. 209–11. Illustrated.

Harmuth, E. Ein Balestrino. *Waffen- und Kostümkunde* (Munich) 3rd series **XIV 1** 1972, pp. 31–4. Illustrated.

Horwitz, H. T. Zwei Konvergenzer-scheinungen in der Waffentechnik. *Zeitschrift für Historische Waffenkunde* (Dresden) **VIII** 1918–20, pp. 171–4. Illustrated. Discusses a Grenade-throwing crossbow.

Hough, W. Note on a Korean Crossbow and Arrow-tube. *American Anthropologist* (Washington DC) New series **I** 1899, p. 200.

Isles, F. W. Anatomy of an Antique (Belgian Crossbow). *Bow and Arrow* (Covina, CA) **III** November 1965, pp. 62–4. Reprinted *Journal of the Society of Archer-Antiquaries* **8** 1965, pp. 31–4.

Lavin, James D. The gift of James I to Felipe III of Spain. *Journal of the Arms and Armour Society* **XIV 2** 1992, pp. 64–88.

Mann, J. G. A Crossbow of Good King René. *Connoisseur* **XCII** April 1934, pp. 235–9.

Millard, G. Pellet-popping Crossbow (made by Conway of Manchester). *TAM & Archery World* (Milwaukee, WI) **XV** February 1966, pp. 34–6. Illustrated.

Morin, M. Armi da Fuoco Particolari Nelle Collezioni dei Musei Italiani: l'Opera di Renaldo de Visin da Asolo. *Diana Armi* (Florence) November/December 1971, pp. 20–3. Illustrated. A crossbow-pistol in the Palazzo Ducale, Venice.

— La Ruota Arcaica. *Diana Armi* (Florence) May/June 1971, pp. 88–91. Illustrated. Three combined crossbows–wheellock guns in the Palazzo Ducale, Venice.

— Una balestrino del XVI Secolo. *Diana Armi* (Florence) June 1977, pp. 32–3. Illustrated.

Murray, A. Note on an Ancient Crossbow Found

Under the Moss on the Estate of Auchmeddan, Aberdeenshire. *Proceedings of the Society of Antiquaries of Scotland* (Edinburgh) **IV** 1860–2, p. 592.

Osmerod, J.G. Crossbow or Prod ? *British Archer* (Portsmouth, Hants) **XIV** 1962–3, p. 146.

Paterson, W. F. The Skein Bow. *Journal of the Society of Archer-Antiquaries* **7** 1964, pp. 24–7.

— The Lion Cock and Crocodile Spanner (French Target Crossbow c.1780). *Journal of the Society of Archer-Antiquaries* (Swalecliffe, Kent) **18** 1975, pp. 38–9.

Picchianti, R. Il Balestrone di Bista. *Diana Armi* (Florence) February 1978, pp. 58–60. Illustrated. The making of a modern crossbow.

Reid, W. The Present of Spain: a Seventeenth Century Royal Gift. *Connoisseur* **CXLVI** August 1960, pp. 22–6. A gift of crossbows sent by James VI and I of Great Britain to Philip III of Spain.

Rieser, R. G. The Elastic Weapons of R. E. Hodges. *Journal of the Society of Archer-Antiquaries* **31** 1988, pp. 13–9. Illustrated.

Rutherford, T. Note on an Iron Man Trap and a Wooden (Whaling) Crossbow Found in an Old Cottage at Seahouses, Northumberland. *Proceedings of the Society of Antiquaries of Newcastle-upon-Tyne* (Newcastle-upon-Tyne) Series 3 **V** 1913, p. 223.

Snoddy, O. An Irish Bullet-crossbow. *Journal of the Arms and Armour Society* **6** 1970, pp. 348–50.

Tucker, W. E. Crossbow Prod. *British Archer* (Portsmouth, Hants) **XIV** 1962–3, p. 193.

Wilson, G. M. Stone Crossbows. *Arms Fair Guide* September 1976, pp. 42–4. Illustrated.

D Bolt and bolt-heads

Anon. Bolt Head of Iron Found at Westminster on the Site of the Houses of Parliament. *Journal of the British Archaeological Association* **VI** 1850, p. 149. Illustrated.

Buttin, C. Les Flèches d'Épreuve et les Armures de Botte Cassée. *Revue Savoisienne* (Annecy) **VI** 1906, pp. 1–8.

— La Flèche des Juges du Camp. *Armes Anciennes* (Geneva) **I** 3 1954, pp. 57–64.

Cederström, R. Beobachtungen an Älteren Bolzen. *Zeitschrift für Historische Waffenkunde* (Dresden) **VI** 1912–4, p. 174.

Elmy, D. The 'Iron-Hearted' Arrow. *Journal of the Society of Archer-Antiquaries* **26** 1983, pp. 29–30. Illustrated.

Haidinger, R. von *Beitrag zur Kenntniss der Bolzen und Pfeilformen vom Beginn der Historischen Zeit bis zur Mitte des XIV Jahrhunderts.* Vienna, Wilhelm Braumüller und Sohn, 1879.

Hart, C. *The Royal Forest.* Oxford, 1966. Appendix V: Quarrels for Crossbows in the Thirteenth Century, pp. 167–72.

Isles, F. W. Effect of Condition, Number and Arrangement of Feathers on the Flight of Crossbow Bolts. *Archers' Magazine* (Boyertown, PA) **IV** August 1955, pp. 13–16.

— A Study in Medieval Crossbow Bolts. *Archers' Magazine* (Boyertown, PA) **IX** September 1960, pp. 36–7. Illustrated. Reprinted *Journal of the Society of Archer-Antiquaries* (Swalecliffe, Kent) **5** 1962, pp. 37–8.

Jessop, Oliver Marc *Medieval arrowheads: a study into the forms of arrowhead that occur throughout the British Isles, using a collection from Dryslwyn Castle in South Wales as a case study.* Unpublished thesis submitted as part of the BA Archaeology course at the Department of Archaeology, Durham University, 1993. Illustrated.

Kalmar, J. Pfeilspitzen als Würdezeichen. *Zeitschrift für Historische Waffen und Kostümkunde* (Berlin) **XV** 1937–9, pp. 218–21. Illustrated.

— Armbrust-Pfeilspitzen als Rangabzeichen. *Folia Archaeologica* (Budapest) **IX** 1957, pp. 153–66. Illustrated.

Lewandowski, M. L'Atelier du Fléchier dans la Tour de Pierre au Château de Legnica. *Fasciculi Archaeologiae Historicae* (Wroclaw) **I** 1986, pp. 49–53. Illustrated.

Lhoest, J. G. Pointes de Flèches et de Carreaux du bas Moyen Age Trouvées dans le Lit de la Meuse à Liège. *Armi Antiche* (Turin) **IX** 1962, pp. 83–92. Illustrated.

Nickel, H. Böhmische Prunkfeilspitzen. *Acta Musei Nationalis Pragae* (Prague) Series A **XXIII** 1969, pp. 101–63. Illustrated.

— Ceremonial Arrowheads from Bohemia. *Metropolitan Museum Journal* (New York) **I** 1968, pp. 61–93. Illustrated.

— Addenda to Ceremonial Arrowheads from Bohemia. *Metropolitan Museum Journal* (New York) **IV** 1971, pp. 179–81. Illustrated.

Paterson, W. F. Venetian Archery Equipment. *Journal of the Society of Archer-Antiquaries* (Swalecliffe, Kent) **4** 1962, pp. 23–4. Illustrations of boltheads and quiver.

Prihoda, R. Zur Typologie und Chronologie

Mittelalterlicher Pfeilspitzen und Armbrustbolzen. *Sudeta* (Leipzig) **VIII** 1932, pp. 32–67. Illustrated.

Strobl, M. Spitz und scharf: Armbrustbolzen vom 13. bis 16. Jahrhundert. *Deutsches Waffen Journal* (Schwabisch Hall) **27** April 1991, pp. 526–9.

Webb, A. John Malemort – Kings Quarreler. The King's 'Great Arsenal' – St. Briavels and the Royal Forest of Dean. *Journal of the Society of Archer-Antiquaries* **31** 1988, pp. 40–6; **32** 1989, pp. 52–8.

Wright, H. F. Research into 'latore' [as a Fletching for Crossbow Bolts]. *Journal of the Society of Archer-Antiquaries* (Colchester, Essex) **1** 1958, p. 45.

E Crossbow parts and accessories

Cederström, R. En Tidig Armborst-spännare. *Livrustkammaren* (Stockholm) **II** 1940–2, pp. 81–3. Illustrated.

Clover, P. American Crossbow Targets. *British Archer* (Portsmouth, Hants) **X** 1958–9, p. 157. Illustrated.

Copin, J. *Les Mécanismes d'Arbalètes.* [Offprint from an unidentified journal acquired about 1962 in Belgium, now in Royal Armouries Library.]

Credland, A. G. Crossbow Remains ... The Crossbow Nut. *Journal of the Society of Archer-Antiquaries* **23** 1980, pp. 12–9. Illustrated.

— Crossbow Remains. 2: the Crossbow Stave (with Notes on Early Spanning Devices). *Journal of the Society of Archer-Antiquaries* **24** 1981, pp. 9–16. Illustrated.

— Crossbow Remains, Part 3. *Journal of the Society of Archer-Antiquaries* **25** 1982, pp. 16–21. Text illustrated.

— A crossbow nut from Stray Farm, Holme-on-Spalding Moor, North Humberside. *Journal of the Society of Archer-Antiquaries* **34** 1991, pp. 7–9. Text illustrated.

Denkstein, V. Pavises of the Bohemian Type. *Acta Musei Nationalis Pragae* (Prague) Series A **XVI** 1962–4, pp. 4–35; **XVIII**, pp. 3–4; **XIX**, pp. 1–34.

Dodds, W. Crossbow Locks. *Journal of the Society of Archer-Antiquaries* (Swalecliffe, Kent) **6** 1963, pp. 33–4. Illustrated.

Dunant, J. Rateliers pour Arbalètes au Château d'Avully? *Institut Suisse d'Armes Anciennes Rapport* **3–4** 1979, pp. 93–5. Text illustrated.

Flewett, W.E. Leonardo da Vinci and the Crossbow Lock. *Journal of the Society of Archer-Antiquaries* **24** 1981, pp. 26–30. Illustrated.

Harmuth, E. Belt Spanners for Crossbows. *Art, Arms and Armour* (Chiasso) 1979–80, pp. 100–7. Illustrated.

— Das Armbrust-Seitenvisier, *Waffen- und Kostümkunde* (Munich) 3rd series **XXI 2** 1979, pp. 159–62. Illustrated.

Hoopes, T.T. A Crossbow-lock of the XVI Century. *Brief Essays in Armor and Arms* (Armor and Arms Club, New York) **II** May 1925.

— The Double Set Trigger, *Miscellany of Arms and Armor, presented by Fellow Members of the Armor and Arms Club to Bashford Dean in honor of his 60th Birthday.* (New York) 1927, pp. 36–40. Illustrated.

— Das Früheste Datierbare Radschloss im Nationalmuseum in München. *Zeitschrift für Historische Waffen- und Kostümkunde* (Berlin) **XIII** 1932–4, p. 224–5. Illustrated.

Horwitz, H.T. Ein Chinesisches Armbrustschloss im Amerikanischen Besitz [in the Boston Museum of Fine Arts]. *Zeitschrift für Historische Waffen- und Kostümkunde* (Berlin) **XI** 1926–8, pp. 286–7. Illustrated.

Kalmar, J. Armbrustspannhacken aus dem XV Jahrhundert. *Archaeologlai Ertesitö* (Budapest) **LXXXIX** 1952, p. 37.

Macgregor, A. Two Antler Crossbow Nuts and Some Notes on the Early Development of the Crossbow. *Proceedings of the Society of Antiquaries of Scotland* (Edinburgh) **107** 1975–7, pp. 317–21. Illustrated.

Murray, K.W. A Crossbow Licence, 1516. *Genealogist* New series **X** 1894, pp. 248–9. Illustrated.

Nickel, H. Der Bolzenkasten des Hans Wagner, Pixnschifter, 1539 [in the Metropolitan Museum of Art, New York]. *Waffen- und Kostümkunde* (Munich) 3rd series **XIII 1** 1971, pp. 26–43. Illustrated.

Paterson, W.F. The Chinese Crossbow Lock. *Journal of the Society of Archer-Antiquaries* (Swalecliffe, Kent) **11** 1968, pp. 24–7.

— A String Puzzle (crossbow string end loops). *Journal of the Society of Archer-Antiquaries* (Swalecliffe, Kent) **11** 1968, pp. 12–13.

— Heavy Crossbow Strings: Bowstring Loop Knots. *Journal of the Society of Archer-Antiquaries* (Swalecliffe, Kent) **19** 1976, pp. 16, 28. Illustrated.

Pierce, F.E. A Study of Rotary Crossbow Locks, *Archer's Magazine* (Boyertown, PA) **IX** November 1955, pp. 21–2. Illustrated.

Reitmaier, P. Goisere 'Rohrlstahel'. *Deutsches Waffen Journal* (Schwäbisch Hall) December 1980, pp. 1694–7. Illustrated.

Rhode, F. Die Abzugvorrichtung der frühen Armbrust und ihre Entwicklung. *Zeitschrift für Historische Waffen- und Kostümkunde* (Berlin) **XIII** 1932–4, pp. 100–2. Illustrated.

Schonberg, A.D. von Setzschilde der Stadt Zwickau aus der Werkstatt eines Schildmachers von Komotau, 1441. *Zeitschrift für Historische Waffen- und Kostümkunde* (Berlin) **XVII** 1943–4, pp. 45–6.

F1 Military use of the crossbow

Anon. Balestre Militari. *Diana Armi* (Florence) January 1984, p. 104. Illustrated. Post World War II military crossbows.

Bishop, W. E. A Military Commission of 1427 [ordering the Dispatch of Crossbowmen from Languedoc to Fight the English]. *Journal of the Society of Archer-Antiquaries* (Swalecliffe, Kent) 6 1963, pp. 18–19.

Boudriot, P. D. L'Arbalète de Guerre. *Gazette des Armes* (Paris) March 1979, pp. 17–23. Illustrated.

Credland, A. G. The Crossbow in Modern Warfare. *Journal of the Society of Archer-Antiquaries* 27 1984, pp. 5–20. Illustrated.

— The Crossbow at War. *Journal of the Society of Archer-Antiquaries* 30 1987, pp. 24–6. Illustrated.

— Big Joe, Little Joe: silent weapons of the 1939–45 war. *Guns Review* (Hebden Bridge) 32 June 1992, pp. 452–5.

Delhomme, P. Les Grenades a Fusil dans l'Armée Française, 1914–18. *Gazette des Armes* (Paris) November 1982, pp. 32–6. Includes an illustration of crossbows being used for grenade launching before the issue of regulation launchers.

Harmuth, E. Zur Leistung der Mittelalterlichen Armbrust. *Waffen- und Kostümkunde* (Munich) 3rd series **XIII** 2 1971, pp. 128–36. Illustrated.

Heer, E. *Armes et Armures au Temps des Guerres de Bourgogne, Grandson 1476.* (Lausanne, Centre d'Histoire) 1976, pp. 170–200. Illustrated.

Holmer, P. L. The Military Crossbow in Yorkist England, 1461–1485. *Journal of the Society of Archer-Antiquaries* 22 1979, pp. 11–16.

Lane, F. C. *The Crossbow in the Nautical Revolution of the Middle Ages, Essays in Honor of Robert L Reynolds.* (Kent, OH) 1969, pp. 161–71.

F2 Use of the crossbow for hunting

Anon. Rook Shooting [using Jackson's Patent Steel Crossbow and 'that of Mr. Bragg of Holborn']. *Sporting Almanack and Oracle of Rural Life* IV 1842, p. 30.

Agostini, S. La Balestra da Caccia, *Diana Armi* (Florence) January 1982, pp. 52–4. Illustrated.

Credland, A. G. More about Bow Traps. *Journal of the Society of Archer-Antiquaries* 26 1983, pp. 31–8. Illustrated.

— The Hunting Crossbow in England, from the Time of the Tudors to the End of the Nineteenth Century. *Journal of the Society of Archer-Antiquaries* 30 1987, pp. 40–60. Illustrated.

— Aquatic Shooting with the Crossbow. *Journal of the Society of Archer-Antiquaries* 31 1988, pp. 20–30. Illustrated.

Guha, K. The Crossbow Trap of the Hos of Singhbhym [Bihar, India]. *Anthropos* (?Vienna) **LXI** 3–6 1966, pp. 871–2.

Haimendorf, C. von Furer Sport Among the Primitive Konyak Nagas [Assam–Burma border]. *Illustrated. London News* 193 24 Dec. 1938, pp. 1192–3. Illustrated. Account of a ceremonial head-hunting dance.

Harmuth, E. Zur Spanischen Jagdarmbrust. *Waffen- und Kostümkunde* (Munich) 24 1 1982, pp. 60–4. Illustrated.

Higson. D. *The Bullet Crossbow, with an Extensive Bibliography on the Arbalist.* (Chorley, Lancashire) 1922. Reprinted 1923.

Horwitz, H.T. Eine Merkwürdige Waffe. Eine Armbrustfalle. *Zeitschrift für Historische Waffen- und Kostümkunde* (Berlin) **XII** 1929–31, pp. 84–5.

Strickland, M. Mr Popplewhit is discovered. *Journal of the Society of Archer-Antiquaries* 39 1996, pp. 16–19. Cartoon sequence of an attempted rook shooting with a crossbow.

F3 Use of the crossbow for target-shooting, including accounts of shooting societies

Ackerman, M. W. R. & R. van Hinte Present-day crossbow shooting in the Netherlands. *Journal of the Society of Archer-Antiquaries* 14 1971, pp. 15–17. Text illustrated.

Anon. Archery in the Netherlands [Popinjay shooting]. *Sporting Magazine* 1 March 1793, pp. 322–3.

Bled, O. Histoire des Arbalètriers de Saint Omer,

Dits «Compagnons ou Chevaliers de Saint Georges». *Bulletin de la Société des Antiquaires de la Morinie* (Saint Omer) **XXII** 1893, p. 327.

Bruges, Gilde des Archers de Saint Sébastien, *Histoire de la Gilde des Archers de Saint Sébastien de la Ville de Bruges.* (Bruges) 1947.

Cartwright, J. D. Use of the Crossbow in England. *Field* **XXIII** 19 March 1983.

Coet, E. Notice sur les Compagnies d'Archers et d'Arbalétriers de la Ville de Roye. *Mémoires de la Société des Antiquaires de Picardie* (Amiens) **XX** 1862, pp. 139.

Delaunay, L. A. *Études sur les Anciennes Compagnies d'Archers, d'Arbalètriers et d'Arquebusiers.* (Paris) 1879.

Derode, V. *Les Ghildes Dunkerquoises, Mémoires de la Société Dunkerquoise* (Dunkirk) V 1959, pp. 278–303.

Dillon, Viscount H. A. L. 'The Order of Shotinge Wt. the Crossbow' by M. Beele. *Journal of the Society for Army Historical Research* **VII** 1928, pp. 185–9, illustrated.

Dini, V. *Dell' Antico uso della Balestra in Gubbio, Sansepolcro, Massa Marittima e nella Repubblica di San Marino.* (Arezzo, Italy) 1963.

Dufour, A. & F. Rabut Notes Pour Servir a l'Histoire des Compagnies de Tir en Savoie. *Memoires de la Société Savoisienne d'Histoire et d'Archéologie* (Chambéry) **XIV** 1873, p. 3.

Feldhaus, F. M. Zur Geschichte der Schießscheiben und Schiessbäume. *Zeitschrift für Historische Waffenkunde* (Dresden) **VIII** 1918–20, pp. 84–6, illustrated.

Fouque, V. *Recherches Historiquess sur les Corporations des Archers, Arbalétriers et des Arquebusiers.* (Paris) 1852.

Francia, A. Crossbow Shooting. *Field* **XVII** 6 April 1861, p. 282.

Frey, G. J. Target crossbow shooting in the United States: the evolution and the revolution. *Journal of the Society of Archer-Antiquaries* **35** 1992, pp. 18–28. Text illustrated.

Gaier, C. Une Illustration du Concours de Tir à l'Arbalète de Gand en 1527. *Le Musée d'Armes* (Liège) **62** 1989, pp. 20–1, text illustrated.

Gunter, R. Zur Geschichte der Schweizerischen Bogenschützengesellschaften. *Schuss und Waffen* (Neudam) **I** 15 December 1907, p. 153.

H[ansard], G. A. Flemish Archery. *Sporting Magazine* **81** December 1832, pp. 161–4.

Janvier, A. O. Notice sur les Anciennes Corporations d'Archers, d'Arbalètriers, de Couleuvriniers et d'Arquebusiers des Villes de Picardie. *Mémoires de la Société des Antiquaires de Picardie* (Amiens) **XIV** 1856, pp. 61–372. Also printed separately (Amiens) 1855.

Meyer, H. Die Armbrust als Sportwaffe. *Deutsches Waffen Journal* (Schwäbisch Hall) **I** December 1965, pp. 32–4.

Perrin, A. Les Moines de la Bazoche, les Abbayes Jeunesse, le Tir du Papegai et les Compagnies de l'Arc, de l'Arbalète, de la Couleuvrine et de l'Arquebuse en Savoie et dans les Pays Anciennement Soumis aux Princes. Maison de Savoie deçà les Monts. *Bulletin de la Société Savoisienne d'Histoire et d'Archéologie* (Chambéry) **VIII** 1864–6, pp. 43–75; **IX**, pp. 1–210; **X**, pp. 241–319, illustrated.

Picchianti, R. ... E si Balestri il Palio. *Diana Armi* (Florence) August 1976, pp. 48–53; September 1976, pp. 42–7, illustrated. An account of the Gubbio crossbow shoot.

Stephanus, M. *Der Armbrustschützen Practica, Paul Brachfield.* 1600. Not seen by the present compiler but cited in Harmuth 1975 (B).

G Manufacture and construction of the crossbow

Booth, A. H. Crossbow Production at the Archbishop's Palace, Trondheim, Norway. *Journal of the Society of Archer-Antiquaries* **39** 1996, pp. 94–100. Text illustrated.

Boothroyd, G. Warrington Gunmakers. *Shooting Times* (Windsor, Berks) 19–25 February 1981, p. 13, text illustrated.

Credland, A. G. Notes on the crossbow spanning bench. *Journal of the Society of Archer-Antiquaries* **33** 1990, pp. 13–22.

Forbes, C. A. Belgian Crossbowmaker. *Field* **XXIII** 2 April 1864

Harmuth, E. Ein Ziehbank in Tirol. *Waffen- und Kostümkunde* (Munich) 3rd series **XI** 1 1969, pp. 33–6, illustrated.

— Armbrustteile im Röntgenbild. *Waffen- und Kostümkunde* (Munich) 3rd series **XIX** 2 1977, pp. 129–36, illustrated.

Heer, E. Beitrage zum Restaurierungsproblem der Armbrust. *Institut Suisse d'Armes Anciennes Rapport* (Grandson) **3–4** 1979, pp. 97–103. Illustrated.

Littler, A. An old Wigan trade: the sporting bullet crossbow and its makers, 1640–1840. *Journal of the Society of Archer-Antiquaries* **34** 1991, pp. 30–41. Text illustrated.

Rohde, F. Über die Zusammensetzung der Spätmittelalterlichen Armbrust, Zeitschrift für

Historischew Waffen- und Kostümkunde. (Berlin) 1940. Translated by T. H. Hamilton: Concerning the Construction of the Crossbow in the Late Middle Ages. *Journal of the Society of Archer-Antiquaries* (Swalecliffe, Kent) **19** 1976, pp. 17–20.

H Crossbows in particular collections and accounts of collections which include crossbows

Beard, C.R. The Robert Lyons Scott Collections of Arms and Armour – Gift to the City of Glasgow. *Connoisseur* **CVII** January 1941, pp. 8–14, illustrated.

Bouillon – Musée Ducal *Belles Armes Anciennes.* (Bouillon) 1971. The Solvay collection, now in the Musée d'Armes, Liège, q.v., includes crossbows.

Bremen – Fockemuseum *Bemerkungen zur Waffensammlung des Focke-Museums in Bremen.* Peter Galperin (Bremen) 1981. Marks, illustrated.

Buttin, F. *Catalogue de la Collection d'Armes Anciennes, Européennes et Orientales, de Charles Buttin.* (Rumilly) 1933. Includes crossbows, items 390–402.

Dean, B. Objects from the William Cruger Pell Collection of Arms. *Bulletin of the Metropolitan Museum of Art* (New York) **II** 1907, p. 8. Crossbow and equipment.

Good, L. T. The Crossbow [in the J. M. Davis Arms and Historical Museum, Claremore]. *Gun Report* **41.5** (October) 1995, p. 57.

Jenkinson, G.P. Bow but a Hewn Stick. *British Archer* (Portsmouth, Hants) **X** 1958-9, pp. 78–9, illustrated.

Joubert, F *Catalogue of the Collection of European Arms and Armour Formed by R L Scott.* (Glasgow) 1924. Three vols. Vol. III, section 5: crossbows. Now Glasgow Museums.

Krieger H.W. *The Collection of Primitive Weapons and Armor of the Philippine Islands in the United States National Museum.* (Washington DC) 1926. Includes crossbows.

Liège – Musée d'Armes *Collection Pierre Solvay, Liège.* Nd, about 1980.

London – Dorchester Hotel *The Loan Exhibition: the Art of the Crossbow.* 1983, pp. 59–69, illustrated.

Mann, Sir J. *Wallace Collection Catalogues: European Arms and Armour: Volume II Arms.* 1962, pp. 477–93, includes marks, crossbows accessories, illustrated.

Markes, O. *Fernwaffen im Wandel der Zeit: Waffen aus der Sammlung Otto Markes.* (Basel, Chur) 1985 Schriftenreihe des Rätischen Museums Chur, 32.

Meyrick, S.R. & J. Skelton *Engraved Illustrations of Antient Arms and Armour, from the Collection of L Meyrick.* 1830.

Milwaukee – Public Museum of the City of Milwaukee *The Rudolph J Nunnemacher Collection of Projectile Arms.* By John Metschl 2 vols (Milwaukee) 1928. Vol. I, Long Arms, includes crossbows.

New York – Museum of Primitive Art *The Lipchitz Collection.* (New York) 1960. Includes crossbows.

Norman, A.V.B. *Wallace Collection Catalogues: European Arms and Armour Supplement* (London) 1986 pp. 202–6.

Paterson, W. F. Manchester Museum: Crossbows in the Simon Collection. *Journal of the Society of Archer-Antiquaries* (Canterbury) **20** 1977, pp. 47–8. Illustrated.

— Uncommon Crossbows from the Simon Collection (Manchester Museum). *Institut Suisse d'Armes Anciennes Rapport* (Grandson) **3–4** 1979, pp. 89–92. Illustrated.

— Traveller's Tales: II: A Castle Armoury. *Journal of the Society of Archer-Antiquaries* **22** 1979, pp. 18–20. Illustrates crossbows, etc., in Burg Kreuzenstein and Kunsthistorisches Museum, Vienna.

Rade, R. *Arbaletele de la Muzeul Militar National, Inchinan sui Nicolae Iorga.* (Cluj) 1931.

Reid, W. A. Royal Crossbow in the Scott Collection. *Scottish Art Review* (Glasgow) **VII** 2 1959, pp. 10–13, 29–30.

Schalkhausser, E. Die Handfeuer-waffen des Bayerischen Nationalmuseums. *Waffen- und Kostümkunde* (Munich) 3rd series **IX** 1 1967, pp. 1–27, includes crossbow with firearms mounted above the stock.

Scott, J. G. *Composite Crossbows in Glasgow Art Gallery and Museum.* Report of the Sixth Congress of the International Association of Museums of Arms and Military History (Zurich) 1972.

Tang, Mei Chun Crossbows of the Formosan Aborigines in the Department Collections. *K'ao ku jen leu hsueh k'an* (Taiwan University, Department of Archaeology and Anthropology, Bulletin) (Taiwan) **VII** 1956, pp. 52–5.

Tarassuk, L. The Collection of Arms and Armour in the State Hermitage, Leningrad,

Journal of the Arms and Armour Society **III** 1959, pp. 1–39. Part I, European Arms, includes crossbows.

Tsarkoe Selo – Musée de Tsarskoe Selo *Musée de Tsarskoe Selo, ou Collection d'Armes de Sa Majesté l'Empereur de toutes les Russies.* (Paris) 1835–53, includes crossbows, bolts, pavises.

Wegeli, R. *Inventar der Waffensammlung des Bernischen Historischen Museums in Bern.* 4 vols (Bern) 1920–48. Vol. IV *Fernwaffen* includes crossbows.

Wilson, G. *Crossbows.* 1976. Treasures of the Tower series, illustrating crossbows in the Royal Armouries.